DOI: https://doi.org/10.52305/LGXP1720

NOTICE TO THE READER

Library of Congress Cataloging-in-Publication Data

ISBN: 979-8-89530-562-1 (Hardcover)
ISBN: 979-8-89530-654-3 (eBook)

Published by Nova Science Publishers, Inc. † New York

Hussain H. Al-Kayiem
Raed A. Jessam
Hasanain A. Abdul Wahhab
Abdul Salam K. Darwish
Editors

Advancements in Sustainability Systems II

nova
science publishers
www.novapublishers.com

Thank you for supporting our efforts to serve you and promote sustainability by reading our book.

Contents

Preface

In a world facing a huge challenge on the environment and climate change, the concept of "sustainability" is marked as a beacon of hope to mitigate the challenges. The book in hand, "*Advancements on Sustainability Systems Development II*," is a continuation of the previous book 'Advancements on Sustainability Systems Development.' Solutions to many global drawbacks are embedded in the adoption of the seventeen identified sustainable development goals (SDGs). The materials selected in the book serve as a crucial guide for professionals, students, and advocates who are passionate about generating resilient systems and methods that advance environmental negative impacts, energy security, and economic development capabilities.

This book compiles materials produced and arranged into 11 chapters that provide important data, information, historical aspects, findings, prospects, and identification of related matters and topics in some of the UNDP identified SDGs. In particular, the techniques and systems for human health and safety, the urban development systems and approaches, and advancements on energy systems for clean power generation have been considered in the book. The Artificial Intelligent as the state-of-the-art approach has also been considered in some chapters, mainly on the environmental monitoring and control. Each chapter in the book draws upon real-world case studies, cutting-edge research outcomes, and the insights of visionaries and practitioners in the aspects related to the SDGs.

As it is not a hidden fact that in addition to the environmental and climate change issues, we witness conflicts up to the military and war levels, as well as trade wars that require us to innovate, develop, and introduce new resolution ideas for resource depletion and social disparities. This book is designed to bridge the gap between theory and practice, providing readers with a comprehensive understanding of sustainability systems development, alongside practical frameworks and strategies for implementation by adoption of state-of-the-art methods.

Chapter 1

The Sustainable Environment in Ajman: A Journey Towards a Green Future in the United Arab Emirates

Ahmed Sebihi[1,*]
Piyush K. Jain[2]
and Rahul Jain[3]
[1]Skyline University College, Sharjah, UAE
[2]Christ University, Bengaluru, India
[3]De Montfort University, Dubai, UAE

Abstract

Ajman, a rapidly developing emirate in the United Arab Emirates, is embarking on a journey toward a sustainable future amidst environmental challenges. This paper explores the current environmental situation in Ajman, analyzing challenges such as pollution, waste management issues, resource depletion, and impacts of climate change. It evaluates ongoing environmental innovation initiatives, including projects addressing these challenges and their impact on the local environment. Additionally, the paper examines the latest environmental technologies and innovations deployed in Ajman, along with efforts to promote environmental awareness among the local community. Furthermore, it delves into partnerships between the government and private sectors and international collaborations to advance sustainability goals. Finally, the paper identifies future challenges and development

---

* Corresponding Author's Email: ahmed.sebihi@skylineuniversity.ac.ae

In: Advancements in Sustainability Systems II
Editors: H. H. Al-Kayiem, R. A. Jessam, Hasanain A. A. Wahhab et al.
ISBN: 979-8-89530-562-1

opportunities, reviews existing environmental policies and legislation, and provides recommendations for fostering a greener and more sustainable future in Ajman.

Keywords: sustainability, environmental challenges, innovation, technology, awareness, partnerships

1. Introduction

Ajman, nestled along the coast of the Arabian Gulf, is one of the seven emirates forming the United Arab Emirates (UAE). With a rich history dating back centuries, Ajman has witnessed remarkable economic and infrastructural development in recent decades (Ajman Department of Economic Development, 2020). This emirate has strategically positioned itself as an emerging economic center within the UAE, capitalizing on tourism, real estate, manufacturing, and trade sectors. The UAE, a federation established in 1971, has rapidly evolved into a global hub for commerce, finance, and tourism. Its visionary leadership has propelled the nation towards unprecedented prosperity, leveraging its abundant oil resources and diversifying its economy (United Arab Emirates Ministry of Cabinet Affairs, 2021). Today, the UAE is a beacon of modernity and progress in the Middle East, attracting millions of visitors and investors annually.

In rapid development, sustainability and environmental conservation cannot be overstated. Like other emirates in the UAE, Ajman confronts pressing ecological challenges stemming from urbanization, industrialization, and population growth. These challenges include air and water pollution, waste management issues, depletion of natural resources, and the looming threat of climate change (Environment Agency - Abu Dhabi, 2019).

Embracing sustainability principles is critical for preserving Ajman's natural heritage, ensuring the well-being of its residents, and safeguarding its economic prosperity. By adopting sustainable practices, the emirate can enhance its resilience to environmental risks, reduce its ecological footprint, and foster a high quality of life for present and future generations (UNDP, 2021).

The purpose of this chapter is to conduct a comprehensive examination of Ajman's journey towards a sustainable future. By analyzing the current state of environmental affairs, evaluating existing initiatives and policies, and exploring avenues for innovation and collaboration, this paper seeks valuable insights into the opportunities and challenges associated with sustainable developments in Ajman. It aims to inform policymakers, stakeholders, and the public about prioritizing environmental sustainability in the emirate's development agenda through rigorous research and analysis.

2. Environmental Challenges in Ajman

2.1. Pollution (Air, Water, and Soil)

Pollution, encompassing air, water, and soil contamination, constitutes a formidable environmental challenge in Ajman. Urbanization, industrial activities, and vehicular emissions contribute significantly to air pollution, particularly in densely populated areas (Environment Agency - Abu Dhabi, 2019). Discharging untreated wastewater and industrial effluents into water bodies such as the Arabian Gulf and local watercourses results in water pollution, jeopardizing aquatic ecosystems and public health (Ajman Sewerage, n.d.). Moreover, soil pollution from improper waste disposal and industrial activities threatens agricultural productivity and ecosystem integrity, necessitating urgent remediation efforts (El Haggar, 2018).

2.2. Waste Management Issues

Ajman grapples with persistent challenges in waste management, exacerbated by rapid urbanization and population growth. Inefficient waste collection systems and inadequate infrastructure contribute to the improper disposal of municipal solid waste, leading to littering, pollution, and public health hazards (Ajman Municipality, 2021). Industrial and hazardous waste generated by manufacturing and construction activities further strain waste management systems, requiring comprehensive strategies for recycling, treatment, and disposal (Al-Salem et al. 2020).

2.3. Depletion of Natural Resources

The depletion of natural resources poses a looming threat to Ajman's environmental sustainability. Groundwater overexploitation, driven by agricultural and urban water demands, results in declining water tables and saltwater intrusion along the coastal aquifers (Environment Agency - Abu Dhabi, 2019). Deforestation, habitat fragmentation, and overfishing exacerbate biodiversity loss and ecosystem degradation, undermining the resilience of Ajman's natural environment (UNEP, 2020).

2.4. Impact of Climate Change

Climate change exacerbates environmental vulnerabilities in Ajman, amplifying the frequency and intensity of extreme weather events. Rising temperatures, shifting precipitation patterns, and sea-level rise heighten the risk of heatwaves, floods, and coastal erosion, posing significant challenges to infrastructure, agriculture, and public health (IPCC, 2021). As a low-lying coastal region, Ajman is particularly vulnerable to the impacts of climate change, necessitating adaptive measures and long-term mitigation strategies (Ministry of Climate Change and Environment, 2020).

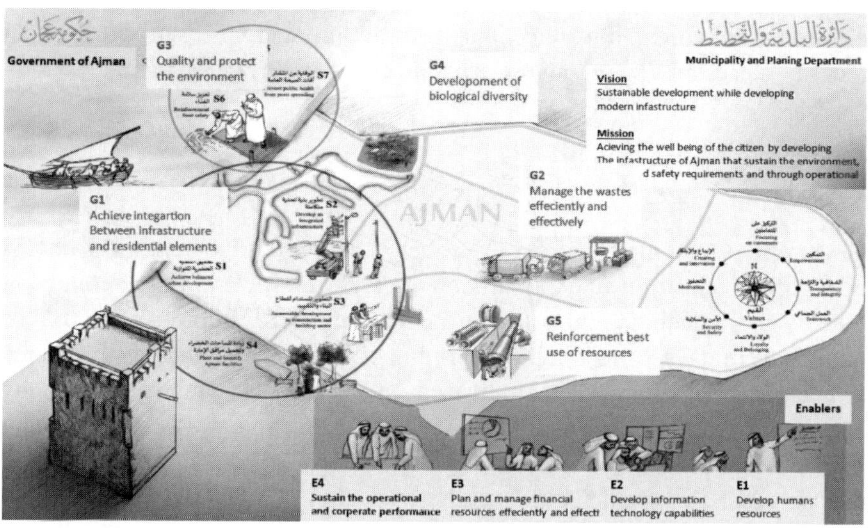

Source: https://www.am.gov.ae/about-us/municipality-strategy/ with permission.

Figure 1. Municipality and Planning Department.

The Municipality & Planning Department of Ajman highlights five strategic goals, shown in Figure 1, aimed at promoting sustainable urban development. These goals include enhancing the quality of life through

- G1: green spaces and recreational areas,
- G2: developing sustainable infrastructure and efficient transportation systems,
- G3: conserving the environment by managing water resources and reducing waste,
- G4: fostering economic growth through eco-friendly businesses and job creation, and
- G5: Engaging the community in sustainability education and awareness programs.

The visual also integrates Ajman's cultural heritage with modern sustainable practices, emphasizing the city's commitment to comprehensive environmental and economic sustainability.

3. Government Initiatives and Policies

3.1. Overview of UAE's Environmental Policies

The United Arab Emirates (UAE) has demonstrated a commitment to environmental sustainability through comprehensive policy frameworks. The UAE Vision 2021, a strategic roadmap for national development, emphasizes sustainable development as a core pillar (Government of Dubai, 2015). The UAE's National Climate Change Plan 2050 outlines ambitious targets for reducing greenhouse gas emissions and enhancing climate resilience (Ministry of Climate Change and Environment, 2020). These initiatives underscore the UAE's dedication to addressing environmental challenges and transitioning towards a greener economy.

3.2. Specific Initiatives in Ajman Aimed at Sustainability

In Ajman, specific initiatives have been implemented to promote sustainability and environmental stewardship. For instance, the Ajman Green Building Regulations mandate environmentally friendly construction practices to reduce energy consumption and carbon emissions (Ajman Municipality, 2021). Furthermore, the Ajman 2021 Strategic Plan includes targets for enhancing environmental sustainability, such as increasing green spaces and improving waste management infrastructure (Ajman Executive Council, 2017). These initiatives reflect Ajman's commitment to integrating sustainability principles into urban planning and development.

3.3. Analysis of the Effectiveness of These Policies and Initiatives

The effectiveness of the UAE's environmental policies and Ajman's sustainability initiatives depends on various factors, including implementation mechanisms, stakeholder engagement, and resource allocation. While progress has been made in advancing sustainability goals, challenges remain in achieving desired outcomes. Factors such as limited enforcement capacity, economic constraints, and social attitudes toward sustainability influence the success of these initiatives (Al-Salem et al. 2020). Therefore, continuous monitoring, evaluation, and stakeholder collaboration are essential to ensure the long-term effectiveness and impact of sustainability efforts in Ajman.

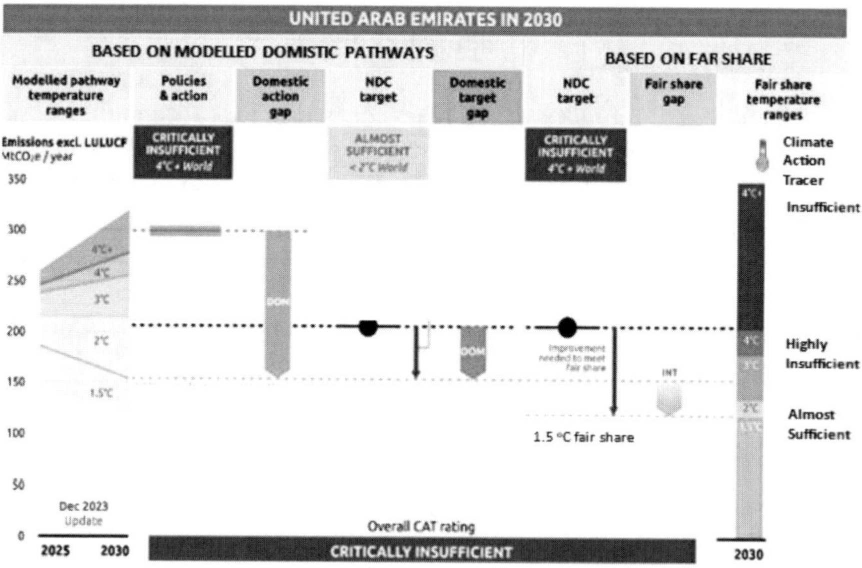

Source: https://climateactiontracker.org/countries/uae/policies-action/ with permission.

Figure 2. Policies and Action by the UAE.

As shown in Figure 2, "The United Arab Emirates in 2030" evaluates the UAE's climate policies and their projected impact on global warming. The analysis shows that the UAE's current policies and actions are rated as "Critically Insufficient," aligning with a pathway leading to over 4°C warming. The Nationally Determined Contributions (NDCs) are rated "Almost Sufficient" for a 2°C world but still "Critically Insufficient" for a 1.5°C world. Significant gaps exist between the current actions, the NDC targets, and the required actions to limit warming to 1.5°C. Additionally, when assessed based on fair share contributions, the UAE's efforts are rated as "Critically Insufficient." The overall Climate Action Tracker (CAT) rating for the UAE is "Critically Insufficient," indicating their current trajectory is inadequate to prevent significant global warming.

4. Green Infrastructure and Urban Planning

4.1. Development of Eco-Friendly Infrastructure

In Ajman, significant efforts have been made towards developing eco-friendly infrastructure as part of sustainable urban planning initiatives. This includes implementing energy-efficient buildings, renewable energy systems, and water conservation measures. For instance, the Ajman Green Building Regulations mandate sustainable construction practices, promoting environmentally friendly materials and energy-efficient designs (Ajman Municipality, 2021). Furthermore, projects such as the Ajman Sustainable City exemplify the integration of green infrastructure, including renewable energy sources and smart technologies, to create environmentally conscious communities (Ajman Government, n.d.).

4.2. Incorporation of Green Spaces in Urban Planning

Integrating green spaces into urban planning is a key focus in Ajman, aiming to enhance biodiversity, mitigate urban heat island effects, and improve residents' overall quality of life. Urban parks, recreational areas, and green corridors are strategically incorporated into city plans to provide residents with access to nature and promote well-being. The Ajman 2021 Strategic Plan prioritizes the development of green spaces and public amenities to create sustainable and livable communities (Ajman Executive Council, 2017). Initiatives like the Ajman Urban Greening Program further emphasize the importance of urban greenery in enhancing environmental sustainability and urban resilience (Ajman Municipality, 2021).

4.3. Case Studies of Sustainable Urban Projects in Ajman

The Al Zorah project is a notable example of sustainable urban development in Ajman. Al Zorah is a waterfront development that prioritizes environmental conservation and sustainable living. It incorporates green building principles, sustainable landscaping, and water management strategies to minimize environmental impact while providing residents with a high-quality living environment (Al Zorah Development Company, n.d.). The Ajman Corniche Development Project also aims to transform the city's waterfront area into a vibrant and sustainable urban space featuring pedestrian-friendly walkways, public parks, and recreational facilities (Ajman Government, n.d.).

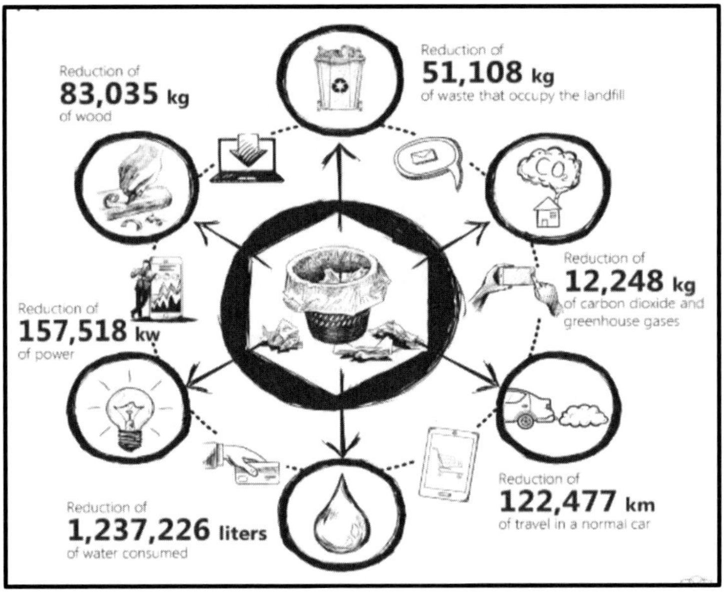

Source: https://www.am.gov.ae/wp-content/uploads/2021/06/ajman-municipality-sustainability-report-en-faw.pdf with permission.

Figure 3. Environmental Impact of Waste Reduction.

As shown in Figure 3, the environmental impact of waste reduction highlights several key metrics. It shows that reducing waste by 51,108 kg can prevent 83,035 kg of wood from being used and save 1,237,226 liters of water. Additionally, it reduces 157,518 kW of power consumption and 12,248 kg of carbon dioxide and other greenhouse gases. This also translates to a reduction equivalent to 122,477 km of travel in a normal car. Each reduction contributes significantly to conserving natural resources, reducing energy use, and lowering greenhouse gas emissions, demonstrating the broad environmental benefits of waste reduction initiatives.

5. Renewable Energy Initiatives

5.1. Adoption of Renewable Energy Sources

In Ajman, there has been a notable adoption of renewable energy sources, particularly solar energy, as part of efforts to diversify the energy mix and reduce dependence on fossil fuels. Solar photovoltaic (PV) installations have gained prominence, with numerous residential, commercial, and industrial facilities harnessing solar power to meet their energy needs. The Ajman 2021 Strategic Plan promotes renewable energy technologies, including solar and wind power, to enhance energy security and sustainability (Ajman Executive Council, 2017). Additionally, initiatives such as the Mohammed bin Rashid Al Maktoum Solar Park, located nearby in Dubai, contribute to the region's renewable energy capacity (Dubai Electricity and Water Authority, n.d.).

5.2. Government Incentives for Renewable Energy Projects

The government of Ajman offers various incentives and support mechanisms to encourage investment in renewable energy projects. These incentives include feed-in tariffs, subsidies, tax benefits, and streamlined permitting processes for renewable energy installations. The Ajman Green Building Regulations also incentivize integrating renewable energy systems like solar panels into new construction and renovation projects (Ajman Municipality, 2021). Furthermore, the UAE federal government provides financial support and incentives for renewable energy projects through initiatives like the

Emirates Green Fund and the UAE Energy Strategy 2050 (Ministry of Energy and Infrastructure, 2020).

5.3. Progress and Challenges in Transitioning to Renewable Energy

The transition to renewable energy in Ajman faces both progress and challenges. On one hand, there has been significant growth in renewable energy capacity, driven by technological advancements, declining costs, and supportive government policies. The development of utility-scale solar projects and distributed solar installations reflects progress toward achieving renewable energy targets (International Renewable Energy Agency, 2020). However, challenges remain, including grid integration issues, intermittency of renewable sources, and regulatory barriers. Additionally, the dominance of fossil fuels in the energy sector and the need for significant upfront investments pose challenges to accelerating the transition to renewable energy (Khatib, 2021).

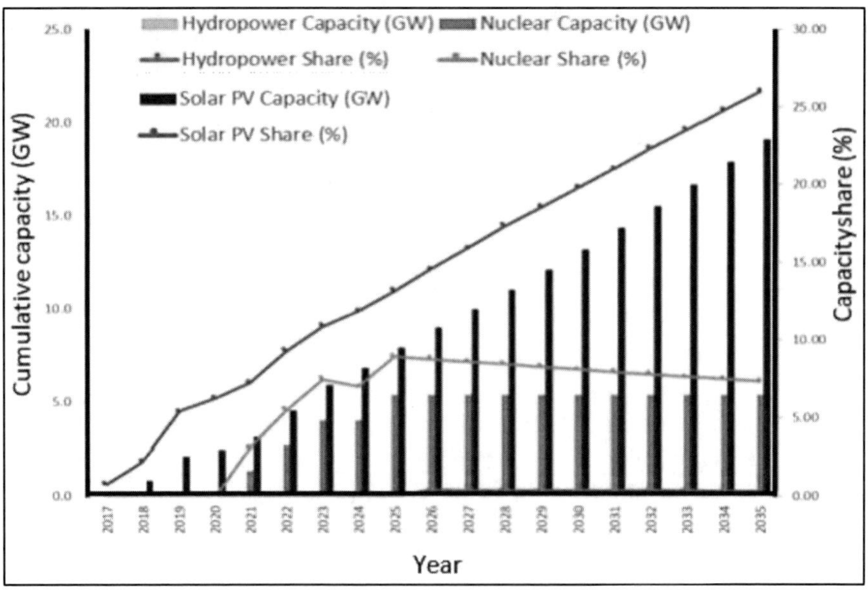

Source: https://www.power-technology.com/wp-content/uploads/sites/21/2022/07/popwer-uae-002.jpg with permission.

Figure 4. Renewable energy to meet clean energy.

The UAE's projected renewable and nuclear energy capacity growth from 2017 to 2035 is displayed in Figure 4. Solar PV capacity (represented in black) shows a significant and continuous increase, expected to reach around 25 GW by 2035, with its share rising to nearly 25%. Hydropower capacity (in teal) remains constant at about 2.5 GW, maintaining a steady share. Nuclear capacity (in orange) will grow to approximately 5 GW by 2025 and then stabilize, with its share declining slightly over time. The trend strongly emphasizes expanding solar PV energy to meet future demands.

6. Public Awareness and Community Engagement

6.1. Educational Programs on Environmental Conservation

In Ajman, various educational programs are implemented to raise awareness of environmental conservation and sustainability. Schools, universities, and community organizations collaborate to integrate environmental topics into curricula and conduct outreach activities. For example, the Ajman Center for Social Responsibility organizes ecological awareness campaigns and school workshops to educate students about waste reduction, water conservation, and biodiversity conservation (Ajman CSR, n.d.). Additionally, initiatives like the Ajman Environmental Education Program provide resources and training for educators to incorporate environmental themes into teaching practices (Ajman Municipality, 2021).

6.2. Community Involvement in Sustainability Efforts

Community involvement plays a crucial role in sustainability efforts in Ajman, with residents actively participating in various initiatives and programs. Local environmental groups, neighborhood associations, and volunteer networks organize clean-up drives, tree-planting events, and awareness campaigns to engage residents in environmental stewardship. For instance, the Ajman Green Team, a community-based organization, mobilizes volunteers to participate in beach clean-ups and community gardening projects (Ajman Green Team, n.d.). Furthermore, initiatives such as the Ajman Community Garden Project encourage residents to cultivate urban gardens and green spaces, fostering a sense of ownership and environmental responsibility among community members (Ajman Municipality, 2021).

6.3. Role of NGOs and Other Organizations in Promoting Environmental Awareness

Non-governmental organizations (NGOs) and civil society organizations significantly promote environmental awareness and advocacy in Ajman. NGOs such as the Emirates Environmental Group (EEG) and the Gulf Research Center engage in research, education, and advocacy initiatives to raise awareness about environmental issues and advocate for sustainable policies and practices (EEG, n.d.; Gulf Research Center, n.d.). These organizations collaborate with government agencies, businesses, and community groups to address environmental challenges, implement sustainability projects, and mobilize public support for environmental conservation efforts.

7. Successes and Challenges

7.1. Assessment of Progress towards a Sustainable Environment

Ajman has made significant strides towards achieving a sustainable environment, evidenced by various initiatives and achievements in environmental conservation and sustainability. Progress can be seen in the increased adoption of renewable energy, the implementation of green infrastructure projects, and heightened public awareness and engagement in environmental issues. For instance, developing eco-friendly infrastructure, such as green buildings and renewable energy installations, reflects the emirate's commitment to reducing carbon emissions and promoting energy efficiency (Ajman Municipality, 2021). Additionally, efforts to enhance waste management practices and preserve natural resources contribute to the overall progress toward a sustainable environment (Al-Salem et al. 2020).

7.2. Identification of Remaining Challenges and Obstacles

Despite progress, Ajman continues to face challenges and obstacles on the path to sustainability. These challenges include air and water pollution, waste management issues, depletion of natural resources, and climate change impacts. Rapid urbanization, population growth, and industrial activities exacerbate environmental pressures, leading to pollution and degradation of ecosystems (Environment Agency - Abu Dhabi, 2019). Additionally, limited resources, technical capacity, and regulatory frameworks challenge the implementation of sustainable solutions and the achievement of desired environmental outcomes (Ajman Executive Council, 2017).

7.3. Comparison with Other Cities or Regions in the UAE or Globally

Compared with other cities or regions in the UAE or globally, Ajman's progress towards sustainability reflects broader trends and challenges urban areas face. While Ajman may lag behind larger cities like Dubai or Abu Dhabi in terms of scale and resources, its efforts towards sustainability align with global sustainability agendas and commitments. Cities such as Masdar City in Abu Dhabi and Dubai Sustainable City serve as examples of ambitious sustainable development projects in the UAE, showcasing innovative solutions for energy efficiency, waste management, and urban planning (Masdar, n.d.; Dubai Sustainable City, n.d.). Globally, cities like Copenhagen, Vancouver, and Singapore are recognized for their comprehensive sustainability initiatives and holistic approaches to urban development, providing valuable lessons and inspiration for Ajman and other cities striving towards sustainability (The World's Most Sustainable Cities, 2021).

8. Future Prospects and Recommendations

8.1. Opportunities for Further Improvement

Despite progress, there are ample opportunities for further improvement in enhancing sustainability in Ajman. Emphasizing green infrastructure development, expanding renewable energy initiatives, and implementing innovative waste management practices present significant avenues for progress. Additionally, fostering partnerships with stakeholders, leveraging technology for environmental monitoring, and enhancing public participation can accelerate sustainability efforts (World Bank, 2020).

8.2. Policy Recommendations for Enhancing Sustainability in Ajman

Policymakers should consider implementing a comprehensive set of policies and regulations to enhance sustainability in Ajman. These may include strengthening environmental protection laws, establishing incentives for renewable energy adoption, and integrating sustainability criteria into urban planning processes. Furthermore, promoting circular economy principles, incentivizing eco-friendly practices among businesses, and investing in sustainable transportation infrastructure can contribute to long-term sustainability goals (United Nations, 2019).

8.3. Predictions for the Future of Environmental Conservation in the Region

Environmental conservation in Ajman is expected to remain a priority, driven by global sustainability imperatives and local development needs. The region will likely witness continued investments in renewable energy projects, green infrastructure development, and conservation initiatives. Moreover, technological advancements, such as smart grids and digital monitoring systems, will enable more efficient resource management and environmental protection. However, addressing climate change impacts, enhancing resilience to natural disasters, and promoting ecological justice will require concerted efforts from governments, businesses, and communities (World Economic Forum, 2021).

9. Moving Beyond Landfills: Ajman's Path to Sustainable Waste Management

Ajman's transition from landfill dependence to a more sustainable waste management system is initially considered by the Ajman Municipality (Ajman

Municipality 21). Environmental risks are associated with landfills, particularly those not properly capped. An important step would be to explore alternatives, such as capping the existing landfill and building a waste-to-energy facility. Using waste-to-energy facilities can reduce the need for fossil fuels and mitigate environmental hazards while diverting waste from landfills. The construction of a waste-to-energy facility can be complex and costly. This type of facility can be expensive to build, maintain, and operate and maintain. Furthermore, obtaining necessary permits and approvals may be challenging, and local communities may oppose the project due to concerns about the impact on the environment (United Arab Emirates Ministry of Cabinet Affairs, 2021).

9.1. Capping and Alternative Solutions for Mitigation of Landfall Impacts

Landfills, especially uncapped ones, significantly threaten the environment. By-products of landfills, such as methane, contribute considerably to global warming. The warming potential of methane is many times greater than carbon dioxide (United Nations Development Program, 2021). A healthier environment for future generations would be achieved by transitioning to a more sustainable waste management system in Ajman, significantly reducing methane emissions (United Nations, 2019). Composting organic waste and capping technologies at landfills will minimize methane production, which should be discussed in this paper section.

9.2. Balancing Community Concerns and the Benefits of Air Quality

Sustainable waste management can benefit from waste-to-energy facilities, but they are not without risks. In the paper, these concerns should be acknowledged, and mitigation strategies should be explored. Consider these key points:

- In the absence of proper management, waste-to-energy plants can emit air pollutants such as nitrogen oxides and particulates. A paper should propose strategies for minimizing air pollution using advanced emission control technologies and rigorous monitoring systems (World Bank, 2020).

- A transparent and community-engaged approach is necessary to address air quality concerns. The paper presents strategies for open communication with residents about the facility's operation and emissions data. Initiatives that engage the community can help build trust and address local anxieties (World Economic Forum, 2021).

Conclusion

Summary of Key Findings

Identified environmental challenges in Ajman: pollution, waste management, and resource depletion.

- Analyzed current initiatives, revealing progress in renewable energy adoption, green infrastructure development, and public awareness campaigns.
- Highlighted persistent challenges, such as the need for ongoing efforts to address sustainability gaps and promote environmental conservation.
- Recognized the importance of understanding the environmental context of Ajman to guide future sustainability efforts effectively.
- Emphasized the significance of data-driven approaches in assessing environmental conditions and tracking progress toward sustainability goals.
- Underlined the role of stakeholder engagement and community participation in driving sustainable change at the local level.
- Noted the importance of government, private sector, civil society, and academia collaboration to implement comprehensive sustainability strategies.
- Identified the need for continuous monitoring and evaluation of sustainability initiatives to ensure effectiveness and identify areas for improvement.
- Acknowledged the diversity of perspectives and expertise required to address complex environmental challenges effectively.

- Stressed the importance of integrating sustainability principles into all decision-making and development planning aspects for a sustainable future.
- The waste-to-energy process presents challenges, while landfills pose environmental hazards. Ajman can benefit from a well-planned transition.
- Waste-to-energy plants need robust emission controls to ensure clean air and community support.

Importance of Continued Efforts Towards Sustainability

- Sustainability is vital for safeguarding the well-being of current and future generations.
- Continued efforts are necessary to mitigate environmental degradation and address pressing sustainability challenges.
- Sustainability promotes resilience to environmental risks, such as climate change, and fosters economic stability and social equity.
- Sustainable practices contribute to preserving natural resources and biodiversity, enhancing the quality of life for all inhabitants.
- Sustainable development aligns with global agendas, such as the United Nations Sustainable Development Goals, driving positive change on a global scale.
- Investing in sustainability creates opportunities for innovation, job creation, and economic growth.
- Sustainable practices contribute to building stronger, more resilient communities capable of facing future challenges.
- Addressing sustainability challenges requires long-term commitment and collaborative action from all sectors of society.
- Sustainable development can help Ajman become a leader in environmental stewardship and responsible governance.
- Embracing sustainability is a moral and strategic necessity for ensuring Ajman's continued prosperity and well-being.

Final Thoughts on the Topic

- Sustainability is an ongoing journey that requires dedication, innovation, and collective action.
- Ajman has made significant strides towards sustainability but must remain vigilant in addressing emerging challenges and opportunities.
- The commitment to sustainability must be ingrained in all governance, planning, and decision-making processes.
- Ajman has the potential to serve as a beacon of sustainable development, inspiring other cities and regions to pursue similar paths.
- By prioritizing sustainability, Ajman can create a legacy of environmental stewardship and resilience for future generations.
- Collaboration and knowledge-sharing with other communities, both locally and globally, can further enrich Ajman's sustainability efforts.
- Every individual has a role in advancing sustainability through small, everyday actions or larger community initiatives.
- Continuous learning, adaptation, and innovation are essential to navigating the complexities of sustainability in a rapidly changing world.
- The journey towards sustainability is not without challenges, but the rewards of a cleaner, healthier, and more prosperous future are well worth the effort.
- As Ajman continues its pursuit of sustainability, let us remain steadfast in our commitment to leaving a positive legacy for generations to come.

References

Ajman CSR. (n.d.). *Ajman Centre for Social Responsibility*. https://www.ajmanchamber.ae/en/about-us/subsidiaries/ajman-social-responsibility-center/#:~:text=The%20Ajman%20CSR%20Center%20is,well%2Dbeing%20in%20the%20emirate.

Ajman Department of Economic Development. (2020). *About Ajman*. https://www.ajmanded.ae/en/about-ajman.

Ajman Executive Council. (2017). *Ajman 2021 Strategic Plan*. [Ajman Executive Council Ajman 2021 Strategic Plan] https://www.ajman.ae/en/node/168.

Ajman Government. (n.d.). *Ajman Sustainable City*. https://www.am.gov.ae/wp-content/uploads/2021/06/ajman-municipality-sustainability-report-en-faw.pdf.

Ajman Green Team. (n.d.). *About Us*. https://www.ajman.ac.ae/en/news/2018/the-aqua-green-team-complete-70-of-their-project-in-the-international-solar-decathlon-competition.

Ajman Municipality. (n.d.). *Waste Management*. [Ajman Municipality Waste Management] https://www.am.gov.ae/wp-content/uploads/2021/06/ajman-municipality-sustainability-report-en-faw.pdf.

Ajman Sewerage. (n.d.). *Environmental Impact Assessment.* [Ajman Sewerage Environmental Impact Assessment] https://www.ajmansewerage.ae/home.

Alhosani, K. M., and Liravi, P. (2022). Sustainable-Business Waste Management: A Case the Emirate of Ajman–UAE. In *Sustainable Energy Development and Innovation* (pp. 447-457). Springer, Cham.

Alhosani, K. M., Kaied, Y. O., Darwish, A. S., & Farrell, P. (2023). Sustainable Waste Management Strategy for Plastic Bag Mitigation and Reducing Carbon Footprint in The Emirate of Ajman–UAE. *Journal of Engineering Science and Technology*, 18(6), 2763-2778.

Alhosani, K. M., Kaied, Y. O., and Darwish, A. S. K. (2021). Ajman an environmentally friendly city with its quality of life: a review of sustainability challenges and achievements by Ajman Municipality and Planning Department. *Renewable energy and environmental sustainability*, 6(12), 1-10.

Al-Salem, S. M., Alsahli, T., & Al-Zahrani, A. A. (2020). Municipal solid waste management in Ajman, United Arab Emirates: Challenges and opportunities. *Waste Management*, 105, 1-12. doi: 10.1016/j.wasman.2020.02.040.

Al-Zorah Development Company. (n.d.). *About Al Zorah*. https://www.alzorahcity.com/about-us.

Dubai Electricity and Water Authority. (n.d.). *Mohammed bin Rashid Al Maktoum Solar Park*. https://www.dewa.gov.ae/en/about-us/strategic-initiatives/mbr-solar-park

Dubai Sustainable City. (n.d.). *About Dubai Sustainable City*. https://www.thesustainablecity.ae/inside-the-city.

EEG. (n.d.). *Emirates Environmental Group*. https://www.eeg-uae.org/.

El Haggar, S. M. (2018). *Sustainable industrial design and waste management: Cradle-to-cradle for sustainable development*. Academic Press.

Environment Agency - Abu Dhabi. (2019). *UAE State of Environment Report 2019.* https://www.ead.gov.ae/en/library/publications/state-of-environment-report.aspx.

Government of Dubai. (2015). *UAE Vision 2021*. [Government of Dubai UAE Vision 2021] https://u.ae/en/about-the-uae/strategies-initiatives-and-awards/strategies-plans-and-visions/strategies-plans-and-visions-untill-2021/vision-2021.

Gulf Research Centre. (n.d.). About Us. https://www.grc.net/.

International Renewable Energy Agency. (2020). United Arab Emirates Renewable Energy Market Analysis. https://www.iea.org/regions/middle-east.

IPCC. (2021). *Climate Change 2021: The Physical Science Basis*. Contribution of Working Group I to the Sixth Assessment Report of the Intergovernmental Panel on Climate Change. Cambridge University Press.

Kaied, Y. O., Darwish, A. S. K., and Farrell, P., 2021. COVID-19 impact on air quality and associated elements: knowledge data of the Emirate of Ajman– UAE. *Renewable Energy and Environmental Sustainability*, 6, p. 15.

Khatib, H. (2021). Renewable Energy Transition in the UAE: Challenges and Opportunities. *Sustainability*, 13(3), 1522. doi:10.3390/su13031522.

Masdar. (n.d.). *About Masdar City*. https://masdarcity.ae/aboutmasdar#:~:text=As%20a%20pioneering%20sustainable%20urban,to%20create%20a%20greener%20tomorrow.

Ministry of Climate Change and Environment. (2020). *Climate Change in UAE*. https://www.moccae.gov.ae/en/about-ministry/irena/about-irena.aspx.

Ministry of Energy and Infrastructure. (2020). *UAE Energy Strategy 2050*. https://www.moei.gov.ae/en/media-center/news/4/7/2023/minister-of-energy-and-infrastructure-reveals-details-of-the-updated-uae-energy-strategy-2050-and#:~:text=He%20added%3A%20%E2%80%9CThe%20updated%20UAE,and%20water%20sectors%20by%202050

Scharff, H., Soon, H. Y., Rwabwehare Taremwa, S., Zegers, D., Dick, B., Villas Bôas Zanon, T. and Shamrock, J., 2023. The impact of landfill management approaches on methane emissions. *Waste Management & Research*, p.0734242X231200742.

The World's Most Sustainable Cities. (2021). *National Geographic*. https://www.nationalgeographic.com/environment/article/worlds-10-most-sustainable-cities-hint-not-fast-growing-jakarta-or-delhi

UNEP. (2020). *Global Biodiversity Outlook 5*. United Nations Environment Program. https://www.unep.org/resources/report/global-biodiversity-outlook-5-gbo-5#:~:text=It%20is%20a%20final%20report,a%20range%20of%20human%20activities.

United Arab Emirates Ministry of Cabinet Affairs. (2021). *About UAE*. https://www.moca.gov.ae/en/about-the-uae.

United Nations Development Program. (2021). *Sustainable Development Goals*. https://www.undp.org/sustainable-development-goals.

United Nations. (2019). *Policy Brief: Circular Economy in Cities*. https://www.un.org/sites/un2.un.org/files/circular_economy_14_march.pdf.

World Bank. (2020). *Green Infrastructure: Smart Investments for our Planet*. https://documents1.worldbank.org/curated/en/343711468343734503/pdf/684910PUB0EPI0067926B09780821395271.pdf.

World Economic Forum. (2021). *The Global Risks Report 2021*. https://www.weforum.org/publications/the-global-risks-report-2021/.

Chapter 2

The Effect of Powder Mixing on the Performance of Electrical Discharge Machining: A Review

Diana A. A. N. Kattab[1]
Zainab H. Mohsein[1]
Laith J. Habeeb[2,*]
and Hasan S. Majdi[3]

[1]Department of Metallurgy and Production Engineering, University of Technology, Baghdad, Iraq
[2]Training and Workshop Center, University of Technology, Baghdad, Iraq
[3]Department of Chemical Engineering and Petroleum Industries, Al-Mustaqbal University College, Hillah, Babylon, Iraq

Abstract

Powder-mixed electric discharge machining (PM-EDM) is a cutting-edge machining technology. This method has gained significant attention due to its ability to overcome the limitations of conventional electric discharge machining (EDM), particularly for high-strength alloys and composites used in industries such as aerospace, automotive, and biomedical applications. By integrating powder into the dielectric fluid, PM-EDM improves the surface quality, material removal rate, and reduces the tool wear rate (TWR). This chapter reviews existing research on the effects of different powders, such as graphite, aluminum, and silicon carbide, on EDM performance. It also discusses the impact of

* Corresponding Author's Email: Laith.J.Habeeb@uotechnology.edu.iq

In: Advancements in Sustainability Systems II
Editors: H. H. Al-Kayiem, R. A. Jessam, Hasanain A. A. Wahhab et al.
ISBN: 979-8-89530-562-1

process parameters like peak current, pulse-on time, and powder concentration on key machining outcomes, including surface roughness and microhardness. The hybrid nature of PM-EDM provides an eco-friendly alternative by minimizing material waste and improving machining efficiency, aligning with sustainability goals in modern manufacturing. Furthermore, by reducing energy consumption through enhanced process efficiency and enabling the machining of advanced materials with higher precision, PM-EDM supports the long-term goal of sustainable production practices. This chapter explores the potential for PM-EDM to advance sustainable manufacturing processes while improving the machining of difficult materials.

Keywords: PM-EDM, performance parameters, powder, electric discharge

1. Introduction

PM-EDM is a cutting-edge machining technology that combines thermal energy from electrical sparks with the introduction of conductive powder particles in a dielectric fluid, offering enhanced performance in the machining of hard-to-cut materials. Ceramics, hardened steels, composites, and superalloys are used in a variety of fields today, including aerospace, missiles, nuclear reactors, and marine industries. For PM-EDM, non-traditional machining processes such as laser machining, EDM, abrasive water jet machining, and plasma machining were used (Singh et al. 2015). Regardless of material hardness, EDM was found to be the best choice for the powder-mixed electrical discharge machining process. Material removal occurs when electrodes are submerged in dielectric fluid because spark energy melts and vaporizes the material from the electrode surface. Applications of the EDM method at greater industrial scales are limited by a lower machining rate, the existence of residual stress, and fracture. Therefore, it is imperative to evolve and create a new machining technique that can both eliminate the shortcomings of the EDM process and manufacture sophisticated materials with the necessary characteristics. PM-EDM is a hybrid machining process developed by researchers that employs powder particles combined with a dielectric fluid in the EDM. Increased corrosion resistance, reduced surface roughness, and higher material removal rate (MRR) are some advantages of using PM-EDM for advanced material machining (Walia et al. 2012).

This chapter discusses PM-EDM, a sustainable machining process that integrates conductive powders into dielectric fluid, improving performance metrics like material removal rate and surface quality while reducing tool wear. This hybrid process aligns with global sustainability goals, facilitating efficient machining of advanced materials in critical industries like aerospace and biomedical. PM-EDM optimizes resources and reduces environmental impact, promoting sustainable manufacturing practices and reducing industrial production's carbon footprint.

2. Background on PM-EDM

PM-EDM evolved as a hybrid technology from traditional EDM, including the technological improvements from its inception in 1946 to the developments in the 1980s, focusing on the introduction of powder into dielectric fluid. Among those who have contributed to the PM-EDM development are Sharma et al. 2010, Sharma et al. 2014, and Bhattacharya et al. 2013. Sharma et al. 2010 used graphite powder in mild steel and claimed enhanced surface finish and material removal rate (MRR). Bhattacharya et al. 2013, through finite element methods, simulated steel machining. He claimed enhanced MRR and heat distribution with powder-based EDM.

Many types of powders are used in dielectric fluid, e.g., graphite, aluminum, and silicon carbide. Each has a different influence on the EDM performance parameters, such as MRR, surface roughness (SR), and tool wear rate (TWR). Kaldhone et al. 2014 studied the machining rates, surface quality, and MRR in PM-EDM of tungsten carbide-cobalt (WC-Co). The effectiveness of the PM-EDM process with tungsten carbide, WC-Co, is evaluated in terms of the MRR. It is observed that copper tungsten is most suitable for use as the tool electrode in EDM of WC-Co. Results revealed that Silicon carbide powder enhances machining rate drastically with a slight increase in tool wear rate TWR. Naif et al. 2015 used aluminum powder mixed in powder Graphite (AISI D2 die steel) in EDM processing. They claimed improved MRR and TWR, and reduced residual stresses and improved MRR with SiC powder.

2.1. Materials Processed by PM-EDM

Different types of materials have been processed in the production and manufacturing process using PM-EDM technology. PM-EDM is capable of machining hard-to-be-machined materials, such as hardened steels, composites, and Inconel alloys. Reported works on the application of PM-EDM processed hard materials are provided in Table 1.

Table 1. Materials processed by PM-EDM

Author(s)	Year	Aim of Study	Conclusions
Ojha et al.	2011	Hastelloy steel	Improved TWR, SR, and MRR using aluminum powder in the dielectric.
Aliyu et al.	2014	EN-8 steel	Enhanced MRR and SR using chromium powder in the dielectric.
Kung et al.	2009	Cobalt-bonded tungsten carbide	Improved MRR and electrode wear using aluminum powder.

2.2. Process Parameters and Their Effects

The input parameters that affect the EDM process are voltage, peak discharge current, spark-on time, duty factor, flushing pressure, workpiece material, tool material, quill-up time, inter-electrode gap, working time, polarity, and powder concentration. Their evaluation as input parameters is their impact on the outcomes, like MRR, TWR, and surface integrity. In addition, the PM-EDM contributes to the surface quality of materials, including parameters like surface roughness, microhardness, and the formation of recast layers. Some related studies are summarized in Table 2.

Table 2. Process parameters and their effects on product surface integrity

Author(s)	Yea	Aim of the study	Conclusions
Sharma et al.	2010	Peak Current	Higher peak current increased MRR but worsened surface roughness.
Bhattacharya et al.	2011	Pulse-On Time	Longer pulse-on time led to deeper craters and higher MRR.
Ekmekci et al.	2016	Surface Roughness (SR) and MRR	Graphite powder improved SR and MRR for the Ti-6Al-4V alloy.
Jabbaripour et al.	2013	Recast Layer Thickness	Reduced recast layer thickness using aluminum powder.
Govindharajan et al.	2014	Surface Topography	Improved topography with nickel-mixed dielectric fluid.

3. PM-EDM Principle

The principle of the PM-EDM is described by Kansal et al. 2005 and Vishwakarma et al. 2012, as shown schematically in Figure 1. Different fine powder materials, such as graphite, aluminum, chromium, copper, and silicon carbide, are mixed into EDM dielectric fluid.

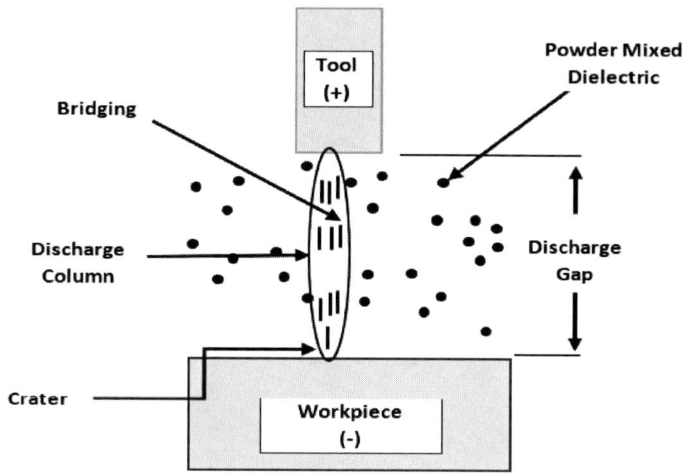

Source: Authors.

Figure 1. Principle of the PM-EDM.

This added metal powder reduces the insulating strength of the dielectric fluid and increases the spark gap distance between the workpiece and electrode tool. Four components make up the removal process in PM-EDM: spark initiation from the fracture of a powder chain, particle zigzag motion from the explosion, and re-bridging of powder particles mixed in the dielectric. When a voltage of 70 – 315 V is applied across the electrodes, an electrical field intensity of 105 V/m is produced. The intensity of the electric field ranges between 105 and 107 V/m.

These particles have been subjected to an electric field due to the impact of the conductive powder particles that are present in the parking area. By bridging the conductors, these conductive powders form a chain-like structure. As a result, an early explosion is caused by a spark gap between the workpiece and the tool. Reduced gap voltage and dielectric strength increase the frequency of discharge and hasten the surface degradation of the workpiece.

The discharge particles spread evenly in all directions, leading the discharge passage to enlarge and widen, resulting in a reduction of the increment in the discharge density and an increase in the discharge heat area (Tripathy et al. 2015), as depicted in Figure 2.

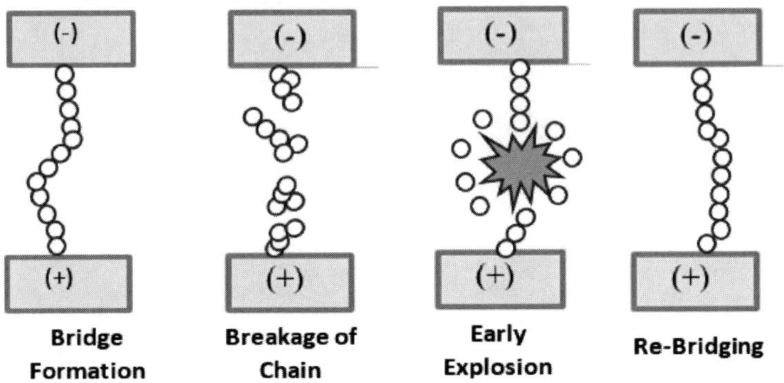

Source: Authors.

Figure 2. The Removal Mechanism of PM-EDM.

4. Experimental Attempts

According to Gugulothu et al. 2015, the arrangement for PM-EDM comprises five components. The dielectric circling framework, which is a fundamental part of the PM-EDM arrangement, contains a flowing siphon and dielectric liquid that flushes the flotsam and jetsam under the cathode hole. It also facilitates recycling powder, cooling the equipment and workpiece, and maintaining the dielectric's temperature below its glimmer point (Kansal et al. 2005). The power supply framework manages the current, voltage, beat frequency, and flash recurrence to create a compelling flash for the material evacuation. The servo-controlled feed framework supports the change of the flash hole between the device and the workpiece, bringing about a legitimate start and steady, effective machining. A mixing framework is comprised of a pulley, a belt, a stirrer, and a DC engine. The stirring system guarantees that powder particles are properly mixed in the dielectric fluid and that powder does not sink at the bottom of the tank (Ekmekci et al. 2016).

The fundamentals of the experimental arrangement of PM-EDM are summarized schematically in Figure 3.

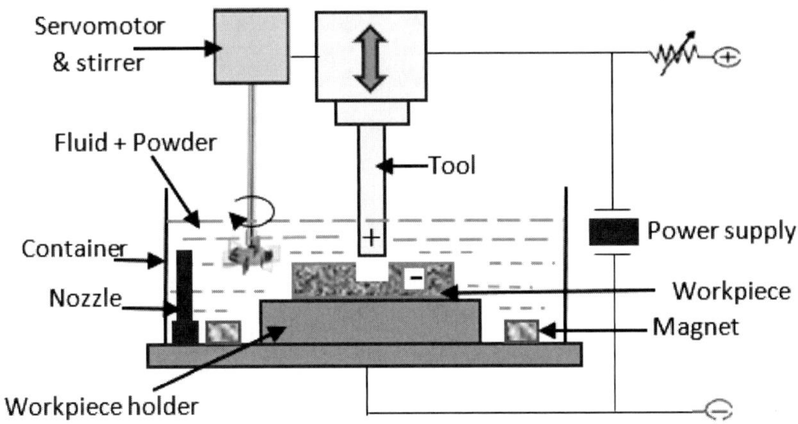

Source: Authors.

Figure 3. Experimental principles of PM-EDM.

According to the authors, the behavior of the particle in the discharge gap can be divided into four categories.

i) Reciprocating.
ii) Stagnation is akin to a chain connecting the electrodes.
iii) Stagnation is akin to a duster in the gap.
iv) Adhesion to either electrode.

Table 3. Properties of various additives powders used in the PM-EDM experiments and their application

Additives	Density g/cm^3	Thermal conductivity (W/m·K)	Electrical conductivity (μΩ/cm)	Melting point (°C)	Specific heat (Cal/g•°C)
Al	2.70	2.38	2.45	660	0.215
Cr	7.16	0.67	12.7	1875	0.11
Sic	3.21	1-5	10^9	2987	0.18
W	19.3	1.673	5.6	3410	0.031
Gr	1.3 – 1.95	0.25 – 4.7	500 – 3,000	4550	0.17 – 0.2
Ni	8.9	0.91	6.84	1453	0.105
Si	2.33	1.5	100,000	1410	0.17
Mo	10.2	1.39	5.27	2610	0.06

Gap voltage and fluid strength are used as dielectrics. Furutani et al. 2001 discussed the types and properties of the powders used in the PM-EDM process. The physical attributes of different powders used in PM-EDM are listed in Table 3.

Different electrode materials have been used as tools in the area listed in Table 4. Various dielectrics that are used as the dielectric fluid in the EDM process are listed in Table 5.

Table 4. Electrode material properties used in the PM-EDM process

Material composition	Graphite	Copper	Copper-tungsten (75% tungsten, 25% copper)
Density (g/cm^2)	1.811	8.904	15.2
Melting point (°C)	3,350	1083	3,500
Electrical resistivity (μΩ.cm)	1400	9	5.5
Hardness	HB 10	HB100	HB 200

Table 5. Characteristics of several dielectric fluids used in EDM

Medium	Properties		
	Dielectric strength (mV/m)	Dielectric strength (mV/m)	Thermal conductivity (W/m·K)
Kerosene	14-22	1.64	0.149
De-ionized water	13	0.92	0.606
EDM Oil	45	2-5	0.954

5. Method of Material Removal in EDM

The material removal process in EDM is illustrated in Figure 4(i) above. During the ignition phase, a large potential difference occurs between the electrodes (Figure 4(ii)). As the electrode moves closer, the electric field increases in the inter-electrode gap until the dielectric breakdown voltage is attained. In the plasma phase, when the currents begin to flow through this phase, as dielectric ionization occurs, the voltage decreases, as shown in Figure 4(iii).

In the discharge phase, the workpiece is continuously heated as the temperature rises, and as a result, there will be a continuous flow of discharge current, forming a molten pool on the electrode while the others directly vaporize. The molten metal pool grows in size, whereas the plasma widens, as shown in Figure 4(iv).

In the ejection phase, when the voltage is turned off, the pressure exerted by the surrounding dielectric decreases, and the plasma channel collapses. The molten is henceforth drawn to the dielectric, Figure 4(v), and removed as a molten metal, and at a later time, it will solidify. This solidified molten is further washed through the inter-electrode gap. The gap thus increases after the spark shifts location.

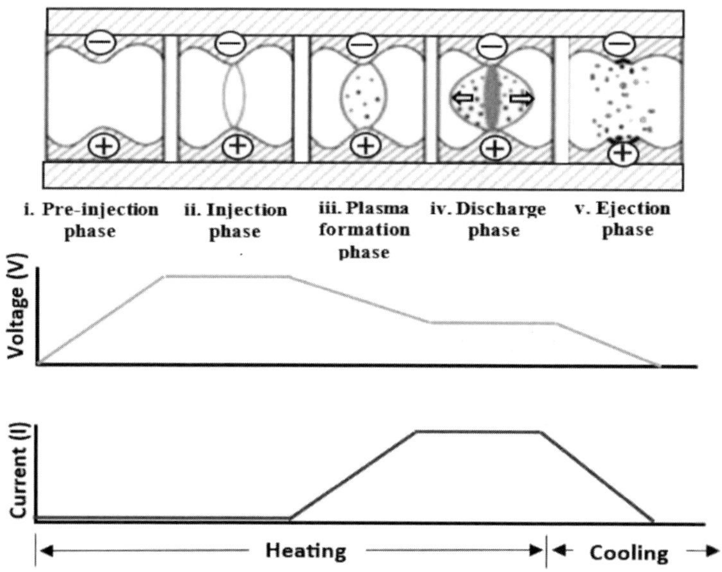

Source: Authors.

Figure 4. Material removal in EDM.

The high temperatures and the cooling cause the re-solidification of the molten metal, resulting in the formation of a recast layer (Bai et al. 2013). Post-machining operations are required to restore the surface integrity (Kansal et al. 2005). Even though the EDM process can be conducted, because of the poor surface quality, its usage is limited. Researchers created an enhanced type of EDM, thus improving its performance. PM-EDM is an example of the EDM process.

6. PM-EDM Process

The PM-EDM process, as shown in Figure 5, is as follows: When an appropriate voltage is applied, an electric field is then generated, causing positive and negative charges to form that move in a specific pattern, that is, a zigzag, and improve the spark gap between electrodes. As the current flows, interlocking also occurs, reducing the insulating strength of the dielectric and increasing the gap.

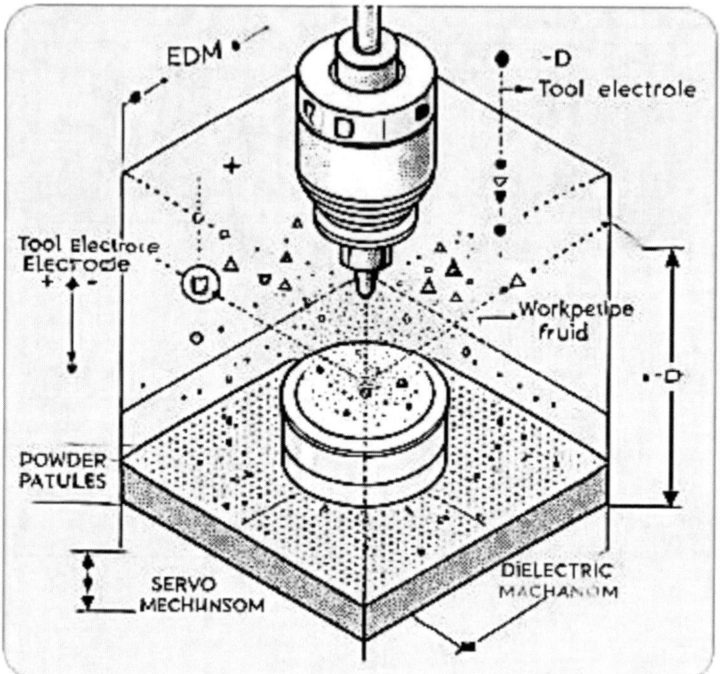

Source: Authors.

Figure 5. Principle of PM-EDM process.

7. PM-EDM Process

The two major categories of PM-EDM are process parameters, as manifested in Figure 6, performance parameters by Choudhary et al. 2014, and gauges, as evinced in Figure 7. To get the best results, the following can be done: enlargement of the discharge gap by Chow et al. 2008, widening of the passage path of the discharge by Kansal et al. 2005, and finally, using multiple discharges by Chow et al. 2008 and Kung et al. 2009.

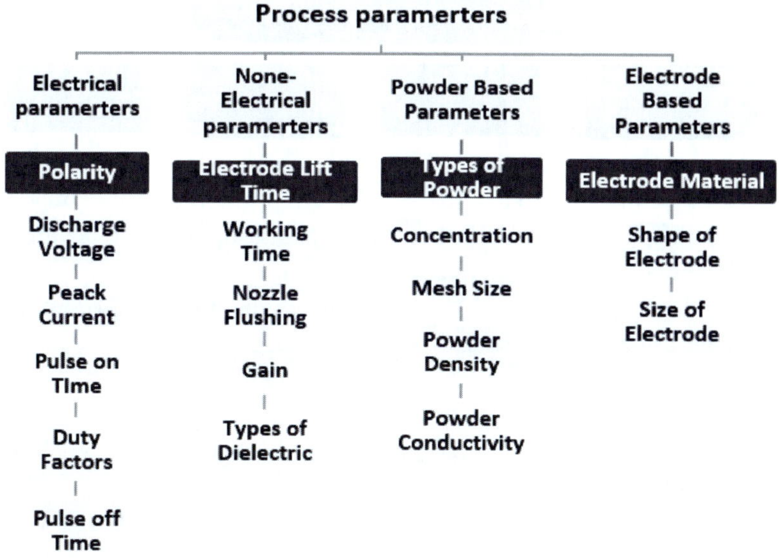

Source: Authors.

Figure 6. Classification of process parameters.

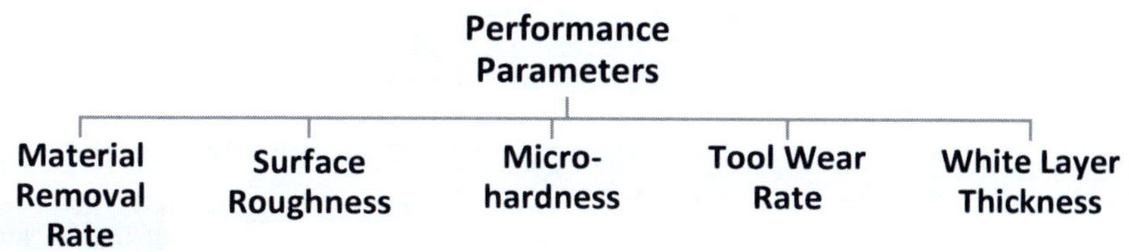

Source: Authors.

Figure 7. Classification of performance parameters.

7.1. Process Parameters

7.1.1. Electrical Parameters

i) Peak current

When a pulse is turned on, the current increases. The amperage requirement is determined by the surface area to be done in the machining processes, the specific impulsive force increases, and the current allows the removal of material (Puertas et al. 2004; Prihandana et al. 2009). As a result, high peak currents and a larger surface area should be used.

ii) Discharge voltage

The voltage increases until the current can pass through an ionization path created by the voltage. The open-circuit voltage drops as the current starts to move through the plasma channel and eventually stabilizes. An adequate voltage is chosen so that there is a working gap between the electrodes. The inter-electrode spacing determines the discharge voltage (Kansal et al. 2005).

iii) Pulse-on time

It is the length of time a stream flows through a cycle. Dielectric ionization occurs during this period. A deeper heat zone, a larger crater, and a larger recast layer are the outcomes of the energy being directed onto the workpiece with a longer pulse on time (Assarzadeh et al. 2013).

iv) Pulse-off time

It is the time taken for the supply voltage to diminish, where the dielectric regains its strength. The time should be maintained at a minimum.

v) Duty factor

It is the ratio of the time spent on the pulse to the total time spent on the cycle. A higher duty cycle results in a higher MRR.

vi) Polarity

The positive polarity, displayed in Table 6 as the optimum combination, is connected to the workpiece, while the negative polarity is connected to the tool. Unwanted material will then be removed from a positive electrode. Longer discharges result in increased tool wear because, as the pulse duration increases, the amount of ion flow increases.

Table 6. Optimum parametric combination in normal polarity

Requirements	Optimal Combination	Confirmation of results		
		MRR (mm^3/min	TWR (mm^3/min	Ra (μm)
Higher MRR	18 A /100 μs/2gm/lit	12.452	0.159	11.105
Lower TWR	6 A /100μs /without powder	2.500	0.056	5.845
Lower Ra	6 A /300 μs/2gm/lit	2.070	0.072	4.818

Table 7. Optimum parametric combination in reverse polarity

Requirements	Optimal Combination	Confirmation of results		
		MRR (mm^3/min	TWR (mm^3/min	Ra (μm)
Higher MRR	18 A / 300 μs/1gm/lit	0.450	0.134	6.881
Lower TWR	6 A / 100 μs /without powder	0.186	0.084	3.155
Lower Ra	6 A / 200 μs/1gm/lit	0.295	0.185	1.655

While reverse polarity, shown in table 7, produces a superior surface polish, normal polarity yields a higher metal removal rate and less tool wear. In the PM-EDM process, normal polarity produces a tool wear rate that is four times lower and an MRR that is 10-21 times greater than reverse polarity. Since reverse polarity offers 1.5-2 times more surface polish than normal polarity, it is preferred for better surface quality (Gavali et al. 2016). The non-electrical parameters, electrode-based parameters, and Powder-based parameters are shown in Table 8.

7.2. Performance Parameters

The most PM-EDM performance evaluation and comparison criteria are summarized and defined below.

7.2.1. Material Removal Rate

It is the volume of material removed per unit of time. From the standpoint of a manufacturer, it is the most important factor. In the PM-EDM, low-melting-point substances have a high MRR (Kiyak et al. 2007). The peak current and pulse influence the MRR greatly. When compared to traditional machining processes, the MRR of the EDM process is low.

Table 8. Parameters of a. non-electrical, b. Electrode-based, and c. Powder-based

a) Non-electrical parameters	
i) Nuzzle flushing	A high dielectric with high strength, Guu et al., 2001.
ii) Working Time	The motion is applied to the workpiece and improves the temperature distribution.
iii) Electrode lifetime	This is the MRR, and this enhances as the vibrating motion improves in the inter-electrode gap flushing Kagaya et al.1986.
iv) Type of dielectric fluid	The following are the functions of the dielectric fluid: (1) It is the insulating medium where the current flows, and this is after the breakdown. (2) it flushes the debris; and (3) It assists in transporting the heat generated by the discharge away from the electrodes.
b) Electrode-based parameters	
i) Electrode material	The electrode should achieve the MMR and have a low tool wear rate, like graphite, which is mostly used to enable high efficiency.
ii) The size and shape of the electrode used	The small size should be used for efficiency.
c) Non-electrical parameters	
i) Type of power	The effects of various powder sources on the process's performance vary. The inorganic oxide does not improve the performance of the dielectric fluid.
ii) Mesh size	The results elucidate that the inter-electrode gap increases with larger particles, resulting in greater contamination, and the lower deionization increases the surface roughness and reduces the MRR Bhattacharya et al. 2013.
iii) Powder conductivity	By increasing the discharge gap with the addition of conductive powder, the frequency and debris flushing are improved. Khosrozadeh et al. 2017.
iv) Concentration	MRR to be high concentration should be increased by increasing powder concentration Sari et al., 2012.
v) Powder density	To achieve a uniform distribution of particles in the dielectric fluid, the low densities help to balance the forces, and as the amount of powder settling decreases, the amount of powder required reduces Schumacher et al. 1983.

7.2.2. Tool Wear Rate

TWR denotes the rate of material evaporation from the tool electrode. It is calculated by dividing the weight loss of the electrode by the time of machining. TWR is critical because it affects the electrode profile, which affects the precision with which the part is machined.

7.2.3. Wear Ratio

WR is the ratio of TWR to MRR. It is an important metric used primarily in engineering, particularly in the fields of manufacturing and materials science, to evaluate the wear performance of materials under specific conditions.

Where:

WR < 1: Indicates that wear rate is lower than material removal rate. This suggests that the material is performing well, with minimal wear under operational conditions.

WR = 1: Indicates a balance between wear and removal, suggesting that the material is worn at a rate like how much material is being removed.

WR > 1: Indicates that the wear rate is higher than the material removal rate. This can be a sign of excessive wear, potentially leading to failure or reduced lifespan of the component.

7.2.4. Surface Roughness

It is defined as the average roughness of the roughness profile's deviations from the central line. Surface roughness increases as heat is focused on the workpiece for a longer period (Tosun et al. 2003). At low operating currents, the surface finish is optimal.

7.2.5. Microhardness

It quantifies the surface roughness, the heated areas, and the recast layer thickness. The fast-cooling, unflushed molten material deposits as a recast layer because of insufficient flushing. The heated zone is a parent material layer beneath the recast layer that has changed the metallurgical properties. The process does not affect the material below this zone. PM-EDM has also been tried as a surface treatment process by researchers (Puertas et al. 2004).

Conclusion

Most of the research uses certain process parameters to investigate their impact on PM-EDM performance attributes such as surface roughness, microhardness, recast layer, and TWR. These factors include pulse-on time, peak current, powder concentration, and pulse-off time. Surface integrity characteristics, including SR, RLT, MH, and surface topography, were improved by adding the required quantity of fine powder to the EDM oil as a dielectric fluid. The conductive material machining method revealed that the most crucial process parameters were peak current, pulse-on time, and powder concentration. In comparison to conventional EDM machining, the machined surface has fewer globular debris particles, voids, and microcracks. The majority of researchers employed Taguchi and RSM methodologies to design trials, evaluate data, and optimize outcomes. A few studies explored soft computing approaches such as gray relational analysis, evolutionary algorithms, and so on when improving experimental findings for multi-response optimization to achieve the required component quality.

Future Scope

Despite the fact that various developments have added to the work on the cutting-edge execution of the PM-EDM and cutting-edge materials, the accompanying work should be possible later on to make hard and conductive materials, for example, molybdenum, high-velocity device preparations, OHNS pass on Inconel, steel, and intermetallic combinations may be machined in different terminal states. Moreover, little is now known about the impact of a number of powders, including titanium, zirconium, manganese, chromium, molybdenum, vanadium, and titanium powder, on high-level material machining. Thirdly, to improve the cycling ability of PM-EDM boundaries and reduce exploratory error, high-level, delicate processes may be applied. It is necessary to conduct a thorough investigation of the influential factors affecting the powder, such as wear resistance and microhardness. Most of the time, the literature describes a dielectric storage tank with a flat bottom. Further research into the impact of different volumes is required to determine the best volume. Investigating the use of PM-EDM in the manufacture of biomedical implants will be beneficial. There has been little research into powder-mixed EDM and alloys, which could also be considered for research and development.

References

Aliyu, A., Musa, H., & Rohani, J. M. (2014). Parametric study of powder mixed electrical discharge machining and mathematical modeling of SiSiC using copper electrode. *Advanced Materials Research, 845,* 878–882. https://doi.org/10.4028/www.scientific.net/AMR.845.878.

Assarzadeh, S., & Ghoreishi, M. (2013). A dual response surface desirability approach to process modeling and optimization of Al_2O_3 powder-mixed electrical discharge machining (PMEDM) parameters. *International Journal of Advanced Manufacturing Technology*. https://doi.org/10.1007/S00170-012-4115-2.

Bai, X., Zhang, Q. H., Yang, T. Y., et al. (2013). Research on material removal rate of powder mixed near dry electrical discharge machining. *International Journal of Advanced Manufacturing Technology, 68(5–8),* 1757–1766. https://doi.org/10.1007/s00170-013-4945-5.

Bhattacharya, A., Batish, A., & Kumar, N. (2013). Surface characterization and material migration during surface modification of die steels with silicon, graphite, and tungsten powder in EDM process. *Journal of Mechanical Science and Technology, 27(1),* 133–140. https://doi.org/10.1007/s12206-012-1142-4.

Bhattacharya, A., Batish, A., & Singh, G. (2011). Optimization of powder mixed electric discharge machining using dummy treated experimental design with analytic hierarchy process. *Journal of Engineering Manufacture,* 103–116.

Choudhary, S. K., & Jadoun, R. S. (2014). Current research issue, trend & applications of powder mixed dielectric electric discharge machining (PMEDM): A review. *International Journal of Engineering Sciences & Research Technology, 3(7),* 335–358.

Chow, H. M., Yang, L. D., Lin, C. T., & Chen, Y. F. (2008). The use of SiC powder in water as dielectric for micro-slit EDM machining. *Journal of Materials Processing Technology, 195(1–3),* 160–170. https://doi.org/10.1016/j.jmatprotec.2007.05.013.

Ekmekci, B., Yaşar, H., & Ekmekci, N. (2016). A discharge separation model for powder mixed electrical discharge machining. *Journal of Manufacturing Science and Engineering, 138(8),* 081006-081006-9. https://doi.org/10.1115/1.4033211.

Furutani, K., Saneto, A., Tabezawa, H., Mohri, N., & Miyabe, H. (2001). Accretions of titanium carbide by electrical discharge machining with powder suspended in working fluid. *Precision Engineering, 25(2),* 138–144.

Gavali, P. B., Waghmare, C. A., & Alashetty, S. S. (2016). Experimental investigation of changing polarity in powder mixed electric discharge machining (PM-EDM). *International Journal of Engineering and Technology, 8(5),* 2049–2058. https://doi.org/10.21817/ijet/2016/v8i5/160805423.

Govindharajan, B., Meivel, P., Chelladurai, C., & Avinaash, K. (2014). Performance and analysis of nickel mixed kerosene servotherm in EDM of Monel 400TM. *Journal of Innovative Research and Solution, 1(1),* 348–354.

Gugulothu, B., Rao, D. H., & Rao, G. K. M. (2015). Optimization of EDM process parameters and graphite powder concentration on electrical discharge machining of Ti6Al-4V alloy using Taguchi method. *International Journal of Advances in Production and Mechanical Engineering, 1,* 31–34.

Guu, Y. H., & Hocheng, H. (2001). Effects of workpiece rotation on machinability during electrical discharge machining. *Journal of Materials and Manufacturing Processes, 16(1),* 91–101. https://doi.org/10.1081/AMP-100103699.

Jabbaripour, B., Sadeghi, M. H., Shabgard, M. R., & Faraji, H. (2013). Investigating surface roughness, material removal rate, and corrosion resistance in PM-EDM of TiAl intermetallic. *Journal of Manufacturing Processes, 15,* 56–68. https://doi.org/10.1016/j.jmapro.2013.01.002.

Kagaya, K., Oishi, Y., & Yada, K. (1986). Micro-electro discharge machining using water as a working fluid-I: micro-hole drilling. *Precision Engineering, 8(3),* 157–162. https://doi.org/10.1016/0141-6359(90)90063-5.

Kaldhone, S. Y., Kavade, M. V., & Rawat, U. (2014). Effect of powder mixed dielectric on performance measures of EDM for tungsten carbide. *International Journal of Innovative Research in Advanced Engineering, 1(10),* 109-111. https://api.semanticscholar.org/CorpusID:212612868.

Kansal, H. K., Singh, S., & Kumar, P. (2005). Parametric optimization of powder mixed electrical discharge machining by response surface methodology. *Journal of Materials Processing Technology, 169,* 427-436. https://doi.org/10.1016/j.jmatprotec.2005.04.051.

Khosrozadeh, B., & Shabgard, M. (2017). Effects of simultaneous ultrasonic vibration of tool and addition of carbon nanotube into the dielectric in EDM process on machining outputs and surface integrity of Ti-6Al-4V alloy. *Indian Journal of Engineering and Materials Science, 24,* 45–56.

Kiyak, M., & Cakır, O. (2007). Examination of machining parameters on surface roughness in EDM of tool steel. *Journal of Materials Processing Technology, 191(1–3),* 141–144. https://doi.org/10.1016/j.jmatprotec.2007.03.037.

Kumar, A., Maheshwari, S., Sharma, C., & Beri, N. (2011). Analysis of machining characteristics in additive mixed electric discharge machining of nickel-based superalloy Inconel 718. *Materials and Manufacturing Processes, 26,* 1011–1018. https://doi.org/10.1080/10426914.2010.538424.

Kung, K. Y., Horng, J. T., & Chiang, K. T. (2009). Material removal rate and electrode wear ratio study on the powder mixed electrical discharge machining of cobalt-bonded tungsten carbide. *International Journal of Advanced Manufacturing Technology, 40,* 95–104. https://doi.org/10.1007/s00170-008-1324-4.

Naif, I., Samir, A. A., & Saad, M. A. (2015). Study the effect of the graphite powder mixing electrical discharge machining on creation of surface residual stresses for AISI D2 die steel using design of experiments. *Journal of Engineering & Technology, 33(A),* 1399–1415.

Ojha, K., Garg, R. K., & Singh, K. K. (2011). Parametric optimization of PM-EDM process using chromium powder mixed dielectric and triangular shape electrodes. *Journal of Minerals & Materials Characterization & Engineering, 10(11),* 1087–1102.

Prihandana, G. S., Maharadika, M., Hamdi, M., & Mitsui, K. (2009). The current methods for improving electrical discharge machining processes. *Recent Patents on Mechanical Engineering, 2(1),* 61–68.

Puertas, I., Luis, C. J., & Alvarez, L. (2004). Analysis of the influence of EDM parameters on surface quality, MRR, and EW of WC–Co. *Journal of Materials Processing Technology, 153,* 1026–1032. https://doi.org/10.1016/j.jmatprotec.2004.06.043.

Sari, M. M., Noordin, M. Y., & Brusa, E. (2012). Evaluating the electrical discharge machining (EDM) parameters with using carbon nanotubes. *IOP Conference Series: Materials Science and Engineering, 40,* 012019. https://doi.org/10.1088/1757-899X/40/1/012019.

Schumacher, B. M. (1983). EDM technology for precision workpieces with excellent surface quality. In *Proceedings of the ISEM 7,* 124–135. https://doi.org/10.1063/1.5123967.

Sharma, R., & Singh, J. (2014). Effect of powder mixed electrical discharge machining (PMEDM) on difficult-to-machine materials – A systematic literature review. *Journal for Manufacturing Science and Production, 14(4),* 233–255. https://doi.org/10.1515/jmsp-2014-0016.

Sharma, S., Kumar, A., Beri, N., & Kumar, D. (2010). Effect of aluminum powder addition in dielectric during electric discharge machining of Hastelloy on machining performance using reverse polarity. *International Journal of Advanced Engineering Technology, 1(3),* 13–24. https://api.semanticscholar.org/CorpusID:58910529.

Singh, A., & Singh, R. (2015). Effect of powder mixed electric discharge machining (PMEDM) on various materials with different powders: A review. *International Journal for Innovative Research in Science & Technology, 2(3),* 164–169. https://ijirst.org/articles/IJIRSTV2I3057.

Tosun, N., Cogun, C., & Inan, A. (2003). The effect of cutting parameters on workpiece surface roughness in wire EDM. *Machining Science and Technology, 7(2),* 209–219. https://doi.org/10.1081/MST-120022688.

Tripathy, S., & Tripathy, D. K. (2016). Multi-attribute optimization of machining process parameters in powder mixed electro-discharge machining using TOPSIS and grey relational analysis. *Engineering Science and Technology, 19,* 62–70. https://doi.org/10.1016/j.jestch.2015.09.008.

Vishwakarma, M., Parashar, V., & Khare, V. K. (2012). Advancement in electric discharge machining on metal matrix composite materials in recent: A review. *International Journal of Scientific and Research Publications, 2(3),* 1–8.

Walia, A. S., Srivastava, V., Garg, M., Somani, N., Gupta, N. K., Prakash, C., Bhargava, C., & Kotecha, K. (2021). Surface roughness analysis of H13 steel during electrical discharge machining process using Cu–TiC sintered electrode. *Materials, 14(20),* 5943. https://doi.org/10.3390/ma14205943.

Chapter 3

Wearable Technologies for Smart Epileptic Seizure Detection

Jehan A. Abbas*
Manaf K. Hussein
and Riyadh A. Abbas
Department of Electrical Engineering, College of Engineering,
Wasit University, Wasit, Iraq

Abstract

Epilepsy affects around 50 million people globally and is marked by unpredictable seizures due to abnormal neural activities in the brain. With nearly 30% of epilepsy patients experiencing drug-resistant seizures, there is a crucial need for effective seizure detection. This paper reviews wearable electronic devices and technologies utilized for smart epileptic seizure detection. These technologies employ various sensors, including accelerometers, gyroscopes, temperature sensors, and photoplethysmography optics, primarily for arm or wrist placement to record seizures. These sensors are valuable for providing accurate seizure data for medical and research use, and they are enhanced with IoT and smartphone integration for real-time data sharing. However, distinguishing between actual seizure movements and normal activities remains a challenge, potentially leading to false positives. Advancements in seizure detection, particularly in multimodal devices, show promise in detecting various seizure types in real-world settings, indicating a positive trend for future innovation and improved patient care. This

* Corresponding Author's Email: Jehan.ali.abbas@gmail.com

In: Advancements in Sustainability Systems II
Editors: H. H. Al-Kayiem, R. A. Jessam, Hasanain A. A. Wahhab et al.
ISBN: 979-8-89530-562-1

review presents the latest updates in seizure detection, including international research and available systems. Wearable devices cover a wide range of seizure types and focus on monitoring autonomic dysfunctions and movement, given their relevance to Sudden Unexpected Death in Epilepsy (SUDEP) and seizure characterization. The development and implementation of these technologies align with the United Nations Sustainable Development Goal (SDG) 3, which aims to ensure healthy lives and promote well-being for all at all ages, highlighting the significance of improving healthcare outcomes for individuals with epilepsy.

Keywords: epileptic seizures, Internet of Things (IoT), non-EEG, smart technologies, wearable device

1. Introduction

The term "Epilepsy" is derived from the Greek verb "epilambanein," which is translated to "to seize" or "to attack." Epilepsy is a neurological condition that presents a significant health challenge (Sheng et al. 2018). Despite the availability of antiepileptic medications, a substantial portion of those affected people, over 30%, suffer from seizures that are resistant to drug therapy (Van de Vel et al. 2014).

The prompt identification of seizure episodes is crucial for summoning aid, which can mitigate the risk of injuries and prevent sudden unexpected death in epilepsy (SUDEP) incidents (van Andel et al. 2016). Epileptic seizures often start with certain changes in the body's autonomic system, such as redness of the skin, increased sweating, and variations in heartbeat rate (Sheng et al. 2018). These changes can potentially be monitored using wearable devices that measure body temperature, skin conductivity (electrodermal activity or EDA), and blood flow changes using a method called "photoplethysmography" (PPG) (Bruno et al. 2018).

Seizures can be of two main types: convulsive and nonconvulsive. Convulsive seizures are characterized by involuntary and repeated muscle contractions and relaxations, leading to noticeable, rhythmic shaking movements. This significant movement during convulsive seizures might be detected through devices that measure movement, such as accelerometers.

On the other hand, nonconvulsive seizures are harder to identify. They may present as brief lapses in attention or a sudden loss of muscle strength, making their detection more challenging. In general, seizure types can be categorized based on their symptoms, as shown in Table 1.

Table 1. Types of epileptic seizures

Seizure Type	Description	Photo
Tonic Seizures (TS)	Characterized by muscle stiffening.	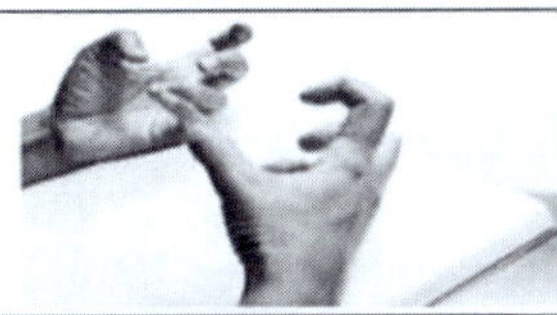
Clonic Seizures (CS)	Involve repeated muscle contractions and relaxations.	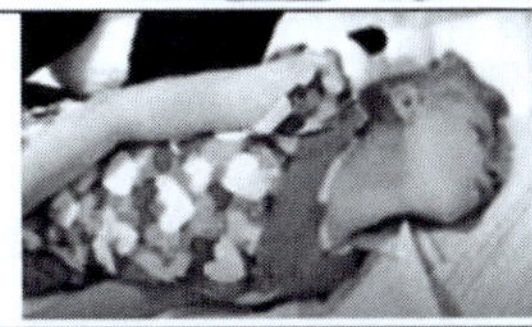
Tonic-clonic Seizures (TCS)	Start with muscle stiffening and then followed by shaking.	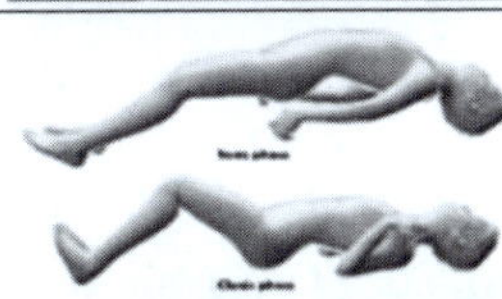
Myoclonic Seizures (MS)	Identified by sudden, brief twitches in muscle groups.	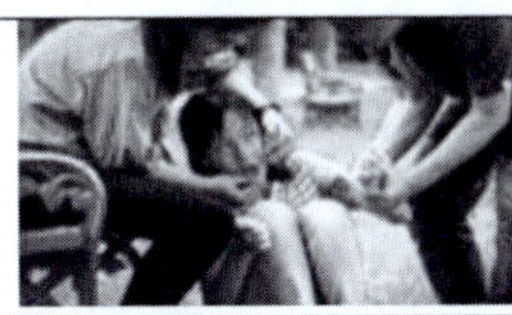
Atonic Seizures	Lead to a sudden loss of muscle strength, often causing falls.	
Absence Seizures	Marked by short periods of unconsciousness, where a person seems unresponsive or unaware of their surroundings, often appearing detached or inattentive.	

Source: Authors.

The treatment and management of epilepsy largely depend on accurately understanding how often and in what way seizures occur. However, relying on patients or their caregivers to report seizures can be unreliable (van Andel et al. 2016). Moreover, keeping a diary of seizures might not always capture the true frequency of these events (Fisher et al. 2012).

An electroencephalogram (EEG) is typically used in medical centers to diagnose and assess seizures. However, there is ongoing research into wearable EEG devices (Bidwell et al. 2015), which could allow for continuous monitoring outside of clinical environments. Despite the significant interest in wearable technologies from those who have epilepsy, their caregivers, and healthcare professionals (van Andel et al. 2016), the options for monitoring seizures during everyday activities and in real-life settings are currently limited to a few wearable seizure-detection devices.

There are also some non-wearable devices, like sensors for beds or mattresses, to detect seizures at night, but this review focuses mainly on wearable technologies. Most patients with epilepsy express a preference for devices that are inconspicuous and can be integrated into commonly used items, such as smartwatches or wristbands. In addition, most wearable devices can be integrated with the Internet of Things (IoT) technology. Such devices represent an array of sophisticated instruments that can be worn, incorporated into clothing, placed inside the body, or even tattooed onto the skin (Firouzi et al. 2023). As shown in Figure 1, various smart wearable devices are designed for epileptic seizure detection. The devices include smartwatches, wristbands, and other IoT-enabled wearables capable of monitoring and recording seizure activities. These devices can connect to the internet, enabling them to send and receive data, which assists in making informed choices. With the evolution of IoT, wearables are moving beyond just being fancy accessories to becoming capable of monitoring and recording seizure activities. These devices can connect to the internet, enabling them to send and receive data, which assists in making informed choices.

With the evolution of IoT, wearables are moving beyond just being fancy accessories to becoming essential devices with specific functions (Kumar et al., n.d.). They are engineered to integrate smoothly with other electronic devices, like smartphones, to boost their operational and communicative abilities, as shown in Figure 2. Given that humans are always on the move, these smart wearables become incredibly useful as they can gather and disseminate data in real-time, and access internet-based information, aiding in making wiser decisions. The adoption of these smart wearables can lead to more effective processes across various sectors, potentially improving overall well-being, productivity, and safety measures (Ghubaish et al. 2021). Advances in low-power communication technologies have accelerated the growth in wearable technology, the reduction in the size of electronic components, and the considerable advantages that these wearables provide (Fahmi Hawari et al. n.d.).

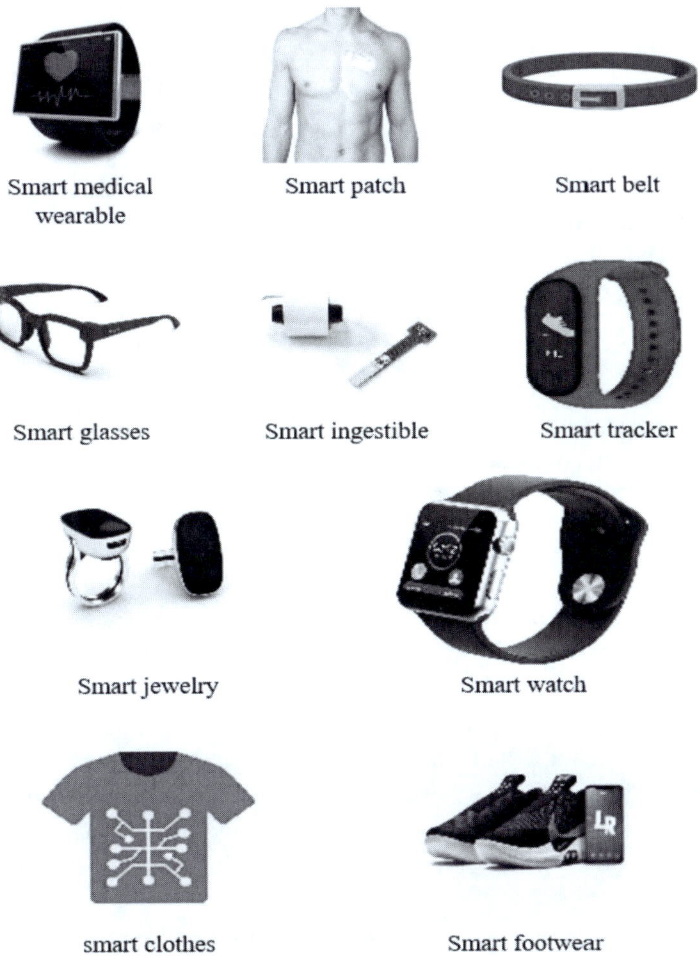

Source: Authors.

Figure 1. Samples of smart wearable devices designed for different uses.

For individuals with epilepsy, wearables designed for monitoring seizures offer significant value. They provide immediate alerts to caregivers when a seizure occurs, reducing the dangers and anxiety associated with such events. These devices could also potentially decrease the threat of SUDEP, which affects a small yet noteworthy number of individuals annually (Islam et al., n.d.). The risk of SUDEP is greater for those living alone or without a roommate, as most incidents occur during sleep and may not be noticed. Rapid response following a seizure could prevent severe consequences or ensure quick medical attention (Wal et al. 2022).

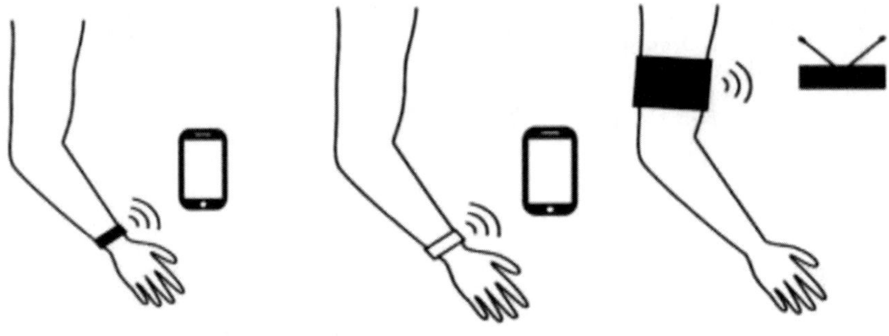

Figure 2. Illustration of Wearable Technologies and Applications for Epileptic Seizure Monitoring: (a) Standalone wrist-based sensing equipment paired with a dedicated application; (b) Software applications that harness data from commercially available wrist trackers; (c) Wearable devices situated outside the wrist region, complemented with an associated base unit (Ghubaish et al. 2021).

While nighttime seizure monitoring devices can help mitigate the risk of SUDEP, evidence regarding their effectiveness in preventing SUDEP outright is still emerging due to the complexity of this area of research. Prompt alerts when seizures happen can also minimize the chances of injury and ensure that emergency medication is given quickly to avoid long seizures (Security et al. 2022). The effectiveness of these devices largely hinges on the presence and reaction of a caregiver. This also can be challenging since many adults with epilepsy live alone, which sometimes leads to social isolation (Ashfaq et al. 2022).

For medical professionals, seizure detection wearables are extremely useful. Self-reported seizures may not be reliable because of the effects of epilepsy itself, medication side effects, or memory issues following a seizure (Lockman et al. 2011). These devices provide a more objective means of monitoring seizures, allowing physicians to better assess how well treatments are working and to make necessary adjustments to medication. Precise data on seizures is also vital for clinical research, particularly when trying to lessen the frequency of seizures (Ding et al. 2023).

This chapter provides a comprehensive review of the existing literature on wearable electronic devices that are designed for seizure detection. This can benefit both individuals with epilepsy and the research community.

2. Sudden Unexpected Death in Epilepsy

A significant number of deaths related to epilepsy are attributed to SUDEP (sudden unexpected death in epilepsy). Although the exact cause of these deaths remains uncertain, various autonomic dysfunctions, including cardiac and respiratory issues, are extensively studied as potential underlying mechanisms. This underscores the importance of incorporating methods to monitor cardiac, respiratory, and other autonomic functions in systems designed for detecting seizures. The topic of SUDEP is sensitive, and due to its unclear mechanisms, many clinicians seldom discuss it with their patients. Discussions are more likely to occur with younger patients, especially those who do not consistently follow their treatment plans. This raises ethical questions about how and when clinicians should introduce the topic of SUDEP to patients and what aspects should be covered. From a patient's perspective, adherence to medication, maintaining a safe sleep environment (like sleeping on one's back and using special pillows), living a healthy lifestyle (avoiding triggers such as cigarettes, alcohol, drugs, and stress), and possibly having supervision at night, are some of the recommended actions to mitigate risks associated with epilepsy Arthurs et al. 2010; Ryvlin et al. 2013).

3. Seizure Detection

Video-EEG is recognized as the gold standard in seizure detection, comprising either scalp-attached electrodes or implanted electrodes for stereotactic EEG purposes.

Implantation, while minimizing EEG artifacts and rendering the system less obtrusive post-installation, carries inherent risks and potential discomfort or social stigma associated with head-mounted devices. There are important questions about how much it costs to set up and maintain devices that predict (Osorio & Manly, 2015). Also, it is necessary to consider if EEG technology can detect all kinds of seizures and if it should be regularly combined with other health data, given the issue of EEG being affected by outside and internal body disturbances, especially when used on the scalp (Bruno et al. 2018). Moreover, there is a question about whether it is possible to stop seizures using medicine or electrical stimulation once they have started, or if the brain has reached a point where the seizure cannot be stopped (Osorio & Manly, 2014).

Further inquiries involve deciding what actions should be taken and at what time to prevent seizures and SUDEP. Determining the most effective brain regions for the localized administration of medication or the application of electrical stimulation is of paramount importance.

4. Seizure Detection Techniques

EEG-based seizure detection is particularly crucial in neonates in Neonatal Intensive Care Units who may have subtle seizures resembling normal behaviors, such as sustained eye-opening, blinking, drooling, or other facial movements (Yamamoto et al. 2011). The distinction between EEG and clinical seizure manifestations can be challenging, and many electrographic seizures do not present clinically. Research in this area includes not just video-based detection (Mathurin et al. 2012) but also ECG (electrocardiography) based approaches (Bidwell et al. 2015; Sheng et al. 2018) and combinations of EEG and ECG (Jeppesen et al. 2015).

Non-EEG sensing devices also play a role. These can be invasive or non-invasive and range from remote sensors like video to contactless methods such as sensor mattresses or e-chairs to wearable technologies. Wearables can be categorized into three types: portable devices like smartwatches or e-accessories; smart textiles like t-shirts or socks with integrated electronic wires or optical fibers capable of measuring various bodily functions; and flexible, stretchable, and printable electronics that can be applied directly to the body as temporary tattoos (Fahmi Hawari et al. 2018).

5. Devices Available in the Market

There are many seizure monitoring devices available for sale in the market worldwide. The best device for someone with epilepsy depends on various factors, including the type of seizures they experience, their age, personal preferences, and what the device does. Generally, there are two kinds of devices:

Table 2. Home-based seizure detection devices on the market are sorted by their mechanisms

Device Name	Detection Mechanism	Placement	Seizure Detection	Availability	Performance Data	Age Range	Price Range
Epi Night Nurse (Wu et al. 2016)	Video Monitoring	Bedside	Clinical Seizures	International	N/A	All Ages	Setup +$395–$495/mo.
Pulse companion (Harden et al. 2017)	HR	Armband	T (Tachycardia)/ B (Bradycardia)	Europe & Australia	N/A	All Ages	$710*
Brio epilepsy monitor (Picard et al. 2017)	HR	Wrist	T/B	UK	N/A	2+ Years	$708*
Pulse Guard Mark II (Myers et al. 2017)	HR	Arm or Ankle	T/B	International	N/A	6+ Months	$511* +$43*/mo.
Epi Care (Elger & Hoppe, 2018)	Motion	Wrist	TC (Tonic-Colinc)	Europe	Good - 90% Sensitivity	10+ Years	Starting at $1985*
Medpage MP5 (Verdru & Van Paesschen, 2020)	BM & Audio	Bed	TC	UK	Variable	25+ kg	Starting at $289*
Sens Alert 200 (Kurada et al. 2019)	BM	Bed	TC	UK	N/A	18+ kg	$955*
Companion Series (Beniczky et al.2021)	BM & Others	Bed	TC	Europe & Australia	N/A	All Ages	$328* to $965*
Nelli home (Simblett et al. 2020)	Video Motion	Bedside	Motor Seizures	Europe	N/A	All Ages	$245*/ day (14–30days)
Sami3 (Halford et al. 2017)	Video Motion	Bedside	TC	US, Australia	N/A	All Ages	$399
Epi Watcher (Verdru & Van Paesschen, 2020)	BM	Bed	TC	International	N/A	All Ages	$1190*
Epi care 3000 (van Westrhenen et al. 2021)	BM	Bed	TC	Europe	N/A	6+ Months	$1504*

Table 2. (Continued)

Device Name	Detection Mechanism	Placement	Seizure Detection	Availability	Performance Data	Age Range	Price Range
Emfit MM Movement Monitor (Jeppesen et al. 2020)	BM	Bed	TC	US, Canada, Mexico	Variable	All Ages	$594
Nightwatch (Geertsema et al. 2018)	Multi-modal	Armband	Various Seizure Types	Europe	Good - 86% Sensitivity	4+ Years	$1822*

* Converted from the original currency to dollars.

Where: HR is Heart Rate, T is Tachycardia, B is Bradycardia, Not available is N/A, TC is Tonic Clonic, and BM is Bed Movement.

Table 3. Portable-based seizure detection devices on the market, sorted by their mechanisms

Device Name	Detection Method	Location Of Detector	Type Of Seizures Targeted	Data On Performance	Suitable Age Group	Notification System	Pricing Information
Epihunter (van Westrhenen et al. 2019)	EEG	Scalp	Absence seizures	Not available	3+ Years	Local Alerts	$48* -$55*/Month
Seizario (Benic zky et al. 2021)	Motion	Waist (Phone Carried)	Absence seizures	Not available	Elderly, kids, adults	Phone Alerts, GPS	Free (needs Android Phone)
EpDetect (Vilel la et al. 2019)	Motion	Waist (Phone Carried)	TC, Fall inducing Seizures	Not available	Elderly, kids, adults	Phone Alerts, GPS	Free (requires smartphone)
SeizAlarm (Si mblett et al. 2020)	Motion/ HR	Wrist or Waist	TC	Not available	Elderly, kids, adults	Phone Alerts, GPS	$14.99/ Month, $149.99/ Year (need iPhone or Apple watch)
Seizure Sync (Geertsem a et al. 2018)	Motion	Wrist	TC	Not available	Elderly, kids, adults	Phone Alerts, GPS	Free (needs Smartwatch & smartphone)
Smart Watch Inspyre (van Westrhenen et al. 2021)	Motion	Wrist	TC	Variable (92.3% sensitivity)	2+ Years	Phone Alerts, GPS	$20+$14.95 - $49.95/ Month (requires Smartwatch & smartphone)
Epi care mobile (Osorio & Manly, 2015)	Motion	Wrist	TC	Good (90% sensitivity)	9+ Years	Phone Alerts, GPS	$1412*
Embrace2 (Lockman et al. 2011)	Multimodal (Motion, Skin Temperature)	Wrist	TC	Good (95% sensitivity)	2+ Years	Phone Alerts, GPS	$249 +$9.90 -$44.90/ Month

Table 3. (Continued)

Device Name	Detection Method	Location Of Detector	Type Of Seizures Targeted	Data On Performance	Suitable Age Group	Notification System	Pricing Information
Seizure Link (Halford et al. 2017)	EMG	Biceps	Motor seizures	Good (93.8% sensitivity)	Arm circumference 6+ in.	Phone Alerts	$499 +~$40 -$45/Month for electrode
SPEAC (Kamiŝ alić et al. 2018)	EMG	Biceps	TC	Good (76% Sensitivity)	17+ Years	Alerts via text, Email, or phone	$4500/ Month for 30–90 days (Prescription required)

* Converted from the original currency to dollars.

i) Home-Based Devices: These are usually set up in one location, such as cameras or bed sensors, and are good for catching seizures that happen at night. They can notify someone in the same house, which is crucial for preventing SUDEP, and they work for people of all ages, including babies (Harden et al. 2017), as illustrated in Table 2.
ii) Portable Devices: These work alongside a smartphone and can be used both at home and outside. They are suitable for individuals who might be alone when they have a seizure because they can alert a contact and provide the user's location (Sveinsson et al. 2020), as shown in Table 3.

Portable devices are helpful for monitoring seizures, especially at night, to prevent SUDEP. They usually integrate with gadgets like phones and smartwatches, which can track movement or heart rate changes. They are typically meant for adults, as they may not be suitable for young children. Although these devices are smart, they can sometimes mistake regular activities for seizures, leading to false alarms, but users can usually switch these off (Geertsema et al. 2018).

One of the main advantages of these devices is that they can accurately record how often seizures occur, which is invaluable for doctors and researchers. For instance, SPEAC, which you can wear, and Nelli, which is used at home, are two devices that help in such a case (Halford et al. 2017). SPEAC can detect muscle movements very precisely, and Nelli can analyze video recordings and differentiate between usual movements and real seizures, with the added feature of sound and video recording. These devices are often used temporarily, for a few weeks to months, to get accurate data on seizure frequency. Nelli has the extra feature of being able to record sound and video, which can help distinguish between other sudden movements and real ones (Verdru & Van Paesschen, 2020).

When choosing the right device, cost, how it looks, and comfort are key factors. Prices vary greatly, from free smartphone apps to expensive devices that can cost over $1000, and some have ongoing fees. Devices that are worn on the body are preferred to be discreet to avoid unwanted attention (Tamura et al. 2014).

Most devices are made to detect tonic-clonic seizures. However, there are some exceptions like Epi-hunter, which is designed for absence seizures detection, and Night Watch, which can detect tonic-clonic seizures. Devices that are designed for tonic-clonic seizures have a high accuracy rate, sometimes over 90%. For example, SPEAC can reach 100% accuracy if it is placed correctly. In contrast, bed sensors like Medpage and Emfit can have lower accuracy. Most devices have fewer than one false alarm per day, but this can vary a lot (Verdru & Van Paesschen, 2020).

For devices that have not undergone rigorous evaluation, the efficacy in detecting tonic-clonic seizures remains ambiguous. Although the underlying technology of these devices may be corroborated by ancillary research, the distinct sensors and algorithms employed can significantly influence their performance. Devices that measure heart rate take advantage of the fact that during seizures, a person's heart rate usually increases. Early studies on devices like ePatch show they are quite accurate and do not have too many false alarms.

Both healthcare providers and researchers need to have devices that can accurately count seizures. Most available devices are set up to spot tonic-clonic seizures. Wearable devices made for these seizures are very accurate. On the other hand, bed-based devices are usually less accurate than wearable devices. Each device has its rate of false alarms, and what is acceptable can vary from person to person (Xu et al. 2019).

For devices that have not undergone expert evaluation, their effectiveness in practical scenarios remains uncertain. Although the foundational concept underpinning these devices is validated, the technology employed may compromise their reliability. Heart rate trackers are useful because most people with epilepsy have a higher heart rate during a seizure. Early tests of ePatch are promising. Night Watch, which uses heart rate and movement tracking, has been tested and works well for different seizure types in real life (Meritam et al. 2018). Other devices have not been able to achieve the accuracy obtained by ePatch and night watch, especially for non-tonic-clonic and motor seizures. To the best of the authors' knowledge, no device has been tested and proven to detect nonconvulsive seizures with high accuracy.

While some nonconvulsive seizures cause noticeable changes in the body, like heart rate, detecting these reliably is hard because of the high number of false alarms (Kurada et al. 2019). However, studies looking back at ePatch show it might be able to detect many nonconvulsive seizures, which gives hope for future research and new devices.

6. Detection Systems

This section summarizes various non-EEG-based methods for detecting seizures, with a focus on non-invasive systems suitable for use as alarm systems in patients' homes. It disregards detection approaches based on metabolic or tissue stress biomarkers due to invasiveness, as well as near-infrared spectroscopy, because it requires wearing a device on the head, which can be conspicuous and potentially lead to stigmatization (van Westrhenen et al. 2021).

6.1. Cardiovascular Changes in Seizure Detection

The identification of distinct patterns or deviations in cardiovascular metrics such as heart rate, blood pressure, and additional cardiovascular indicators associated with, or preceding, seizure events is central to this research. The observation of these variations may facilitate the invention of technologies or methods designed to inform caregivers or patients about the likelihood of forthcoming seizures or to supply valuable data for clinical assessment and the formulation of therapeutic strategies.

6.1.1. Heart Rate Monitoring in Seizure Detection

Heart rate monitoring can be performed using several technologies as below:

- A ballistocardiograph (ECG) captures the electrical activities of the heart and can assess heart rate, heart rate variability, and ECG morphology. Alternative techniques are predominantly focused on heart rate measurement.
- (BCG) monitors the subtle movements of the torso resulting from blood flow. Plethysmography is used to measure changes in volume within organs, applicable for tracking air in respiratory functions or blood in the context of heart rate, blood pressure, and oxygen saturation (Jeppesen et al. 2015).
- Photoplethysmography (PPG) employs LED light to gauge blood volume fluctuations by observing light that is either reflected from or transmitted through the body. PPG is advantageous due to its non-reliance on electrodes, reducing skin irritation risks, and maintaining functionality despite skin contact issues (Teohari et al. 2019). However, (PPG) may be influenced by external lighting and patient movement, although this can be offset by adding motion detection capabilities.

- Phonocardiography (PCG), conversely, monitors the sound of the heartbeat, functioning like a stethoscope (Osorio & Manly, 2015).

An elevated heart rate persisting after exercise is associated with reduced heart rate variability, indicative of autonomic nervous system functioning; a higher degree of variability signifies enhanced adaptability. Furthermore, deviations detected in ballistocardiograph measurements have been recognized as prognostic markers for sudden cardiac death across various medical scenarios (van Andel et al. 2015), reductive indicators of sudden cardiac death in diverse medical conditions.

Seizure-related heart rate changes, especially in generalized tonic-clonic seizures, hyperkinetic frontal lobe seizures, and temporal lobe seizures, are common. These changes can include ictal tachycardia, bradycardia, or asystole, with tachycardia often occurring even before the seizure onset (Jansen et al. 2013), increasing SUDEP risk.

Epilepsy patients typically exhibit reduced heart rate elevation inter-ictal, with heart rate elevation differentiating focal epileptic seizures from physical exercise (Jeppesen et al. 2015). Temporal lobe seizure patients show distinct heart rate elevation patterns compared to psychogenic non-epileptic seizures (Renevey et al. 2001).

Cardiac-based detection systems face challenges in differentiating between various physical states, such as arousal from sleep or physical activities from seizures. A solution could involve integrating a chest accelerometer to determine body position.

Notable developments include a Dutch-Belgian team and the Holst Center/imec and Hobo Heeze BV collaboration, creating a chest and arm-attached system using a ballistocardiograph for seizure detection (Masse et al. 2013). This evolved into a multimodal approach combining ballistocardiograph, accelerometer, video, and audio (Schulc et al. 2011). The Livassured Night Watch, using both an accelerometer and Phonocardiography for heart rate detection, is another development.

In the US, RTI International and the French company Bioserenity's Neuronaute smart textile incorporate ballistocardiographs in their devices. The Emfit seizure-detection system by Finnish company Emfit Ltd. utilizes BCG. LivaNova in the US developed the ProGuardian and AspireSR systems, the latter being an implanted device for ballistocardiograph-based detection with closed-loop vagal nerve stimulation Sabesan & Sankar, 2015).

Photoplethysmography is used in many smart and fitness watches for heart rate detection and is featured in devices by (Cogan et al. 2015; Li et al.

2022), the PulseGuard smartwatch by Adris Technologies, and the Embrace by Empatica (Patterson et al. 2015).

6.1.2. Blood Pressure Measurement and Monitoring

Blood pressure assessment can be conducted via arm cuff devices or finger plethysmography, commonly integrated into pulse oximeters. Finger plethysmography serves to evaluate heart rate and the amplitude of the pulse wave and, when combined with a ballistocardiograph, can determine pulse transit time, which bears an inverse relationship to blood pressure levels (Doyle et al. 2010; Greene et al. 2007). Research on human subjects with drug-induced seizures has documented substantial elevations in blood pressure, with average increases in systolic and diastolic pressures by 82 mmHg and 42 mmHg, respectively. The maximal blood pressure levels are usually reached within one minute following the onset of a seizure, and typically, blood pressure normalizes to baseline values within 60 minutes, regardless of the continuation of status epilepticus (Simon et al. 2010).

6.1.3. Oxygen Saturation Assessment

Pulse oximetry is typically conducted using a clip-like device affixed to the earlobe or fingertip and can be seamlessly incorporated into accessories such as rings or gloves for ease of use. In the case of infants, it is often integrated into a foot strap or a body-adhesive sticker. Research by Cogan et al. 2015 integrated into a finger cuff connected to a watch, enabling the concurrent monitoring of heart rate, oxygen saturation, and electrodermal activity (EDA), which tracks sweating for seizure detection (refer to Table 4).

Research observations suggest that during a seizure, alterations in heart rate occur initially, followed by changes in oxygen saturation and subsequently in EDA. A significant benefit of this approach is the stability of its signal and the simplicity of its detection algorithm, which is activated by a mere decline in the oxygen saturation level. Notably, a reduction of 4% in saturation over 8 seconds is established as the threshold for apnea detection.

6.2. Respiration Changes in Seizure Detection

Changes in respiratory patterns are frequently observed during seizures and are integral in averting SUDEP (Sudden Unexpected Death in Epilepsy). It is essential to track various respiratory behaviors such as apnea, sighing, yawning, and arousal responses. Diminished arousal capacity may serve as an early indicator of imminent SUDEP risk, while arousal and gasping mechanisms play a critical role in spontaneous resuscitation (Poets et al. 2004). Multiple techniques exist for monitoring respiration, including the measurement of airflow, temperature, pressure, or velocity; the observation of chest and abdominal movements; the assessment of blood oxygen levels or respiratory gases; and the application of ECG.

Devices such as thermocouples or thermistors positioned beneath the nose are employed to gauge the temperature of the airflow. Masks that envelop the nose and mouth or pneumotachograph masks are utilized to measure airflow pressure or velocity. Although these devices are commonly used, they can cause discomfort and potentially interfere with natural breathing patterns (Binks et al. 2006).

The measurement of respiratory rate, depth, and effort can be achieved through the detection of chest and abdominal wall movements. This is accomplished using respiratory inductance plethysmography (RIP), which involves two straps, or impedance pneumography, which utilizes two electrodes. Additionally, magnetometers can also be employed for this purpose (Teohari et al. 2019). Single electromyography (EMG) or accelerometer (ACM) sensors, as well as remote sensing techniques using video or microwave technology, are also viable methods for respiratory monitoring. However, these approaches may continue to register respiratory movements even during episodes of apnea. Bed-mounted seizure detection systems such as Ep-It P139, Aremco, and Emfit are capable of sensing respiratory alterations by tracking movement.

Oximetry plays a pivotal role as it can detect elevations in blood pressure resulting from airway obstructions, notwithstanding potentially deceptive respiratory movement indications. Comprehending the interplay between neurological, cardiac, and respiratory systems is critical; therefore, data on blood oxygenation is essential, complementing information on respiration and heart rate (Poets et al. 2004).

Transcutaneous sensors measuring the partial pressure of oxygen offer more precise detection of respiratory anomalies with a reduced incidence of false positives compared to oximeters. In a study by Poets (67), the integration of these sensors with pulse oximetry and chest movement detection has yielded significant insights into the diverse respiratory patterns associated with various medical conditions. Techniques like oxygraphy and capnography, which are employed to quantify the concentrations of oxygen and carbon dioxide in respiratory gases, leverage infrared waves (akin to the technology used in pulse oximetry) or electrodes for their measurements. This combination of methodologies enhances the understanding of respiratory dynamics across different health scenarios (Arthurs et al. 2010).

Instruments such as the compact device developed by Ervitech, which is affixed to the suprasternal notch of the neck, are capable of measuring airflow by identifying sounds emanating from respiratory turbulence. Although this device is theorized to be effective in detecting apnea incidents during seizures, current research has not extensively explored its application in the context of epilepsy (Sabesan & Sankar, 2015).

Electrocardiogram-derived respiration (EDR) represents an alternative approach wherein respiratory activity influences the ECG signal through variations in lung volume. While EDR can identify a disassociation between cardiac rhythm and respiratory patterns during seizures, its relatively slow signal response and the often brief and variable nature of seizures render it less optimal for seizure detection in comparison to methods primarily focused on cardiac activity (Jansen et al. 2013).

6.3. Measuring Skin Conductivity and Other Body Responses

Autonomic responses affecting the skin are present in several forms. Skin flow, as evidenced by skin reddening, can be quantified with electrode-based measurements. Nonetheless, for reactions such as goosebumps or piloerections, there are currently established detection techniques beyond visual observation or self-reporting methods (van Andel et al. 2015). Electrodermal activity (EDA), representing variations in skin resistance or conductance, can be quantified using instruments like ohmmeters or galvanometers, and it signifies the activity of sudomotor (sweat gland) functions.

During generalized tonic-clonic seizures (GTCS), there is a significant escalation in autonomic nervous system discharges, primarily from the sympathetic branch. While physiological parameters such as blood pressure, heart rate variability (HRV), and respiration are regulated by both the parasympathetic and sympathetic nervous systems, the sweat glands are primarily influenced by sympathetic innervation. Consequently, fluctuations in EDA distinctly reflect sympathetic nervous system responses and can be effectively used in the detection of GTCS. However, EDA changes are typically slower to manifest compared to heart rate changes, as noted by Cogan et al. 2015.

Table 4. Research on non-EEG seizure detection approaches, covering development, algorithms, and outcomes

Authors	Method of Detection	No. of Patients	No. of Seizures and Type	Connection	Results
Becq et al. 2011	accelerometer, magnetometer	2	46 TC, C (clonic)	Connect to the wrist	Sensitivity 90%, FDR = 0.7/night
Cogan et al. 2015	photoplethysmography, blood oxygenation, electrodermal activity	5	12 seizures (GTCS), (FS)	Watch, finger cuff	Sensitivity 58%, FDR = 0.01/hour
Conradsen et al. 2012	accelerometer, gyroscope, electromyography	10 healthy, 1 patient	1 real TC (Tonic-Clonic)	Sensor suit, electrodes on the body	Sensitivity 100%, FDR 0/hour, Latency: 0.75s
Gubbi et al. n.d.)	accelerometer	8	14 unspecified seizures	iPod Touch on wrists	Sensitivity 100%, Few FP
(Jallon, 2010	accelerometer	2	46 unspecified seizures	Connect to the wrist	Sensitivity 88-89%, PPV: 55-75%, FDR = 0.5-0.7/night
Masse et al. 2013, van Elmpt et al. 2006	electrocardiography	17	104 (Tonic and TC)	Necklace or arm device and chest electrodes	HR changes in 48% of 8 patients, Sensitivity >90%, PPV >50%
Milošević et al., n.d.)	accelerometer, electromyography	7	31 TC seizures	Connect to limbs	Best ACM result: Sensitivity 86.36%, FDR 1.94/12h
Van de Vel et al. 2013	accelerometer	7	51 (hours of which 2 per patient used)	Connect to limbs	Sensitivity 95.71%, PPV 57.84% in 180h
Beniczky et al. 2013 Epi-Care free	accelerometer	73 (20 with seizures)	39 GTCS	Bracelet	Sensitivity 89.7%, FDR 0.2/day
Carlson et al. 2009 MP5	Audio	64	8 TC	Between bed and a sleeping pillow	Sensitivity 63%, PPV 3.3%
Conradsen et al. Eddi Van Westrhenen et al. 2019	electromyography	5 (2 with seizures)	7 GTCS	Electrodes on the leg	Sensitivity 57%, Latency 25s, FDR 0.003/h

Authors	Method of Detection	No. of Patients	No. of Seizures and Type	Connection	Results
Lockman et al. 2011 Smartwatch	accelerometer	40 (6 with seizures)	8 TC	Watch	Sensitivity 87.5%, PPV >50%
Narechania et al. 2013 Emfit	accelerometer	51 (13 with seizures)	18 GTCS	Sleeping pillow	Sensitivity 89%
Patterson et al. 2015 SmartWatch	accelerometer	41	51 GTCS	Watch	Sensitivity 16-34% for all seizures, High FDR

Where: FDR: False Detection rate; FP: False Positive; PPV: Positive Predictive Value.

Regarding seizure detection, a research team from the United States has documented the development of a wrist-based device known as 'Embrace' by Empatica. This device combines EDA monitoring with an accelerometer (ACM) function. Additionally, it is equipped to measure body temperature and heart rate through photoplethysmography (PPG), with the primary objective of detecting seizures (as detailed in Table 4) (Patterson et al. 2015).

Furthermore, the research team, headed by Cogan et al. 2015, has incorporated EDA into their watch-based design. Both this group and Empatica assert that their devices can detect nonconvulsive focal seizures that involve loss of consciousness, formerly categorized as complex partial seizures. However, the efficacy demonstrated by Cogan's team is relatively modest, and the claims made by Empatica are pending confirmation through comprehensive scientific scrutiny.

6.4. Motor Activity Assessment

This assessment encompasses a diverse array of evaluations and observations aimed at measuring motor abilities, muscular strength, coordination, movement capacity, and proficiency in executing daily activities. It serves multiple purposes, including diagnosis, monitoring of disease evolution, assessment of treatment or intervention efficacy, and the formulation of rehabilitation strategies.

6.4.1. Video Analysis in Seizure Detection

Video analysis is a crucial component of the gold standard in seizure detection, offering the unique capability for retrospective visual analysis of seizure incidents. This method is generally non-invasive, barring instances where infrared wave reflectors are attached to specific parts of the patient's body, like joints or extremities, to enhance motion tracking (Rémi et al. 2011). Video technology can also be potentially employed to assess heart rate variability and respiratory rate (Looy et al. 2011).

A significant challenge in video detection is its limited ability to detect movements under blankets, although the blanket's contours can provide some indications. In some cases, patients might be required to sleep without a blanket or use colored pajamas to facilitate movement detection (Jansen et al. 2013). Additionally, comprehensive movement recording necessitates the patient remaining within the camera's field of view, which is less problematic during nighttime.

Thermographic cameras, which detect infrared emissions from body movements even through blankets and clothing, offer an alternative detection method. These devices have the potential to monitor respiration and heart rate as well (Murthy et al. 2009). However, they are currently disadvantaged by a lower resolution compared to standard optical cameras and a higher cost.

The primary concern with video-based seizure detection methods is privacy. Although the transmitted data may be restricted to alerts or processed signals without including actual visual footage, privacy considerations remain paramount.

Independent research groups in Belgium, the United States, Germany/Portugal, and Italy are advancing in the field of seizure detection via automated video analysis (Mathurin et al. 2012; Murthy et al. 2009; Rémi et al. 2011). The SAMi Alert system, developed by the American company SAMi, employs video technology specifically for the detection of convulsive seizures.

6.4.2. Electromagnetic Wave-Based Movement Detection

The utilization of electromagnetic waves, notably microwaves and radio waves, has been investigated for their capacity to detect movement, heart rate, respiration, and temperature, capitalizing on the Doppler effect (Suzuki et al. 2009). Furthermore, the potential of infrared motion sensors in identifying nocturnal seizures has been explored, as demonstrated in research conducted by Shankar et al. 2013.

These electromagnetic techniques are advantageous as they enable contactless monitoring, capable of penetrating blankets and clothing. However, there are notable drawbacks. One primary concern is the current reliability and accuracy of these devices. Another significant issue is the continuous exposure to electromagnetic radiation.

The potential for radiation damage varies based on the type of wave used. Although radio waves, infrared, and microwaves generally lack ionizing capabilities, they can exert electrical or biological influences on the human body, including thermal effects. The potential hazard linked to radiation exposure depends on various factors, including the intensity of the waves, the cumulative exposure period, and the total duration of radiation exposure.

6.4.3. Movement Analysis in Seizure Detection

ACMs are engineered to quantify translational acceleration and are known for their cost-effectiveness and minimal power requirements. These attributes enable the use of these devices for mobile monitoring, with their compact design, substantial storage capacity, and fast processing capabilities. They are thus well-suited for various medical applications, such as recognizing physical activity patterns. For example, in Parkinson's disease, accelerometers are instrumental in distinguishing between typical movements and motor dysfunctions like hypokinesia, bradykinesia, or dyskinesia (Becq et al. 2011). Additionally, they have proven effective in identifying clonic seizures (Lockman et al. 2011).

Gyroscopes, which measure angular or rotational acceleration, are particularly adept at detecting seizures, although they have higher energy consumption than accelerometers. Conversely, magnetometers evaluate positional and orientational changes, making them useful in detecting tonic seizures characterized by specific postural changes. These devices are also valuable for early detection of the tonic phase in tonic-clonic seizures (Becq et al. 2011). However, magnetometers are susceptible to environmental influences and can be affected by external magnetic fields.

When affixed directly to a patient, these sensors provide a direct measure of movement, including during motor seizures, and can distinguish between the movements of different limbs. It is important to note that if a single accelerometer is used, it must be correctly positioned on the limb consistently involved in the seizures. Alternatively, when attached to a bed or mattress, these sensors are more comfortable for the patient and less likely to be displaced over time.

6.5. Audio Monitoring in Seizure Detection

During seizures, audio surveillance can capture a range of sounds, including typical screams, humming, involuntary laughter or weeping, bronchial secretions, lip-smacking, and the noise of beds caused by patient movement. Moreover, specific audio monitoring devices can detect sounds associated with respiration.

The primary advantages of audio monitoring systems include their affordability, non-invasive nature, and ease of use. These qualities make them popular, especially in pediatric settings, where they are often employed in forms like rattling bracelets or modified baby monitors.

However, audio systems often suffer from poor accuracy and a high rate of false positives. A significant challenge is developing algorithms capable of filtering out background noises and patient-generated sounds like talking or snoring. However, in some instances, snoring may indicate post-seizure activity, such as stertorous breathing following tonic-clonic seizures (Shankar et al. 2013).

Research groups from the United Kingdom and the United States have conducted evaluations on diverse audio monitoring systems, including those developed by Medpage. Comprehensive information about these studies can be found in Table 3. Additionally, a research team from the Netherlands has investigated the use of audio-based seizure detection techniques in combination with other modalities (Poets et al. 2004).

6.6. Eye Movement Detection

Electrooculography (EOG) can capture movements of both the eyelids and the eyes, proving useful in distinguishing epileptic seizures from conditions such as syncope and psychogenic or other types of non-epileptic seizures (Chung et al. 2006). The Epicall system, developed by Epicall Ltd., utilizes a sticker applied to the face for tracking eye movements, heart rate, and pulse, serving as potential indicators of an impending seizure or early seizure activity.

6.7. Temperature Monitoring

Exploration continues into the efficacy of temperature monitoring as a tool for seizure detection. Its application is particularly considered for discerning febrile seizures, although the precise relationship between these seizures and epilepsy is not always definitive. Methods for measuring temperature variations include an array of thermometers, ranging from adhesive stickers and watch probes to radiometers and thermal cameras, with the latter also able to detect movement (Greene et al. 2007).

The role of temperature monitoring is further examined in relation to the analysis of changes in exhaled air, as detailed in Section 6.2. Notably, devices developed by companies such as Empatica and RTI International incorporate temperature measurement capabilities.

Conclusion

In this review, significant strides in wearable technologies for epilepsy have been explored, which is a field that holds immense promise for enhancing patient care and quality of life. These innovative devices, with their array of sensors and integration of IoT, are pivotal in enabling precise monitoring and rapid response to seizures, particularly in drug-resistant cases.

Despite their transformative potential, challenges such as ensuring accuracy, reducing false positives, and addressing the societal stigma of epilepsy remain. Future directions should focus on refining these devices for individualized seizure patterns, exploring non-EEG detection methods, and addressing public perception to encourage broader adoption. The field stands at a crucial juncture, with the potential for these technologies to revolutionize epilepsy management and significantly improve patient outcomes. The need for continued research and development is paramount, as is the pursuit of solutions that are both effective and sensitive to the unique needs of individuals with epilepsy.

Acknowledgments

The authors acknowledge the use of AI tools for proofreading and rephrasing during the writing process.

References

Arthurs, S., Zaveri, H. P., Frei, M. G., & Osorio, I. (2010). Patient and caregiver perspectives on seizure prediction. *Epilepsy & Behavior*, 19(3), 474–477. https://doi.org/10.1016/j.yebeh.2010.08.010.

Ashfaq, Z., Rafay, A., Mumtaz, R., Hassan Zaidi, S. M., Saleem, H., Raza Zaidi, S. A., Mumtaz, S., & Haque, A. (2022). A review of enabling technologies for Internet of Medical Things (IoMT) Ecosystem. *Ain Shams Engineering Journal*, 13(4), 101660. https://doi.org/10.1049/pbhe034e_ch1.

Becq, G., Bonnet, S., Minotti, L., Antonakios, M., Guillemaud, R., & Kahane, P. (2011). Classification of epileptic motor manifestations using inertial and magnetic sensors. *Computers in Biology and Medicine*, 41(1), 46–55. https://doi.org/10.1016/j.compbiomed.2010.11.005.

Beniczky, S., Polster, T., Kjaer, T. W., & Hjalgrim, H. (2013). Detection of generalized tonic–clonic seizures by a wireless wrist accelerometer: A prospective, multicenter study. *Epilepsia*, 54(4), e58–e61. https://doi.org/10.1111/epi.12120.

Beniczky, S., Wiebe, S., Jeppesen, J., Tatum, W. O., Brazdil, M., Wang, Y., Herman, S. T., & Ryvlin, P. (2021). Automated seizure detection using wearable devices: A clinical practice guideline of the International League Against Epilepsy and the International Federation of Clinical Neurophysiology. *Epilepsia*, 62(3), 632–646. https://doi.org/10.1016/j.clinph.2020.12.009.

Bidwell, J., Khuwatsamrit, T., Askew, B., Ehrenberg, J. A., & Helmers, S. (2015). Seizure reporting technologies for epilepsy treatment: A review of clinical information needs and supporting technologies. *Seizure*, 32, 109–117. https://doi.org/10.1016/j.seizure.2015.09.006.

Binks, A. P., Banzett, R. B., & Duvivier, C. (2006). An inexpensive, MRI compatible device to measure tidal volume from chest-wall circumference. *Physiological Measurement,* 28(2), 149. https://doi.org/10.1088/0967-3334/28/2/004.

Bruno, E., Simblett, S., Lang, A., Biondi, A., Odoi, C., Schulze-Bonhage, A., Wykes, T., & Richardson, M. P. (2018). Wearable technology in epilepsy: The views of patients, caregivers, and healthcare professionals. *Epilepsy & Behavior: E&B*, 85, 141–149. https://doi.org/10.1016/j.yebeh.2018.05.044.

Chung, S. S., Gerber, P., & Kirlin, K. A. (2006). *Ictal eye closure is a reliable indicator for psychogenic nonepileptic seizures. Neurology, 66*(11), 1730–1731. https://doi.org/10.1212/01.wnl.0000259662.87864.f9.

Cogan, D., Nourani, M., Harvey, J., & Nagaraddi, V. (2015). *Epileptic seizure detection using wristworn biosensors. Proceedings of the Annual International Conference of the IEEE Engineering in Medicine and Biology Society, EMBS*, 2015-November, 5086–5089. https://doi.org/10.1109/embc.2015.7319535.

Conradsen, I., Beniczky, S., Wolf, P., Kjaer, T. W., Sams, T., & Sorensen, H. B. D. (2012). Automatic multimodal intelligent seizure acquisition (MISA) system for detection of motor seizures from electromyographic data and motion data. *Computer Methods and Programs in Biomedicine, 107*(2), 97–110. https://doi.org/10.1016/j.cmpb.2011.06.005.

Ding, W., Abdel-Basset, M., Hawash, H., Abdel-Razek, S., & Liu, C. (2023). Fed-ESD: Federated learning for efficient epileptic seizure detection in the fog-assisted internet of medical things. *Information Sciences, 630*, 403–419. https://doi.org/10.1016/j.ins.2023.02.052.

Doyle, O. M., Temko, A., Marnane, W., Lightbody, G., & Boylan, G. B. (2010). Heart rate based automatic seizure detection in the newborn. *Medical Engineering & Physics, 32*(8), 829–839. https://doi.org/10.1016/j.medengphy.2010.05.010.

Elger, C. E., & Hoppe, C. (2018). Diagnostic challenges in epilepsy: seizure under-reporting and seizure detection. *The Lancet Neurology, 17*(3), 279–288. https://doi.org/10.1016/s1474-4422(18)30038-3.

Fahmi Hawari, H., Yuhao, L., David Chung Hua, B., Fahmi, H., Che Kiong, C., & Harun, A. (2018). Internet of Things (IOT) monitoring system for elderly. *2018 International Conference on Intelligent and Advanced Systems*, IEEE. https://doi.org/10.20944/preprints202406.0302.v1.

Firouzi, F., Jiang, S., Chakrabarty, K., Farahani, B., Daneshmand, M., Song, J., & Mankodiya, K. (2023). Fusion of IoT, AI, Edge-Fog-Cloud, and Blockchain: Challenges, Solutions, and a Case Study in Healthcare and Medicine. *IEEE Internet of Things Journal, 10*(5), 3686–3705. https://doi.org/10.1109/jiot.2022.3191881.

Fisher, R. S., Blum, D. E., DiVentura, B., Vannest, J., Hixson, J. D., Moss, R., Herman, S. T., Fureman, B. E., & French, J. A. (2012). Seizure diaries for clinical research and practice: limitations and future prospects. *Epilepsy & Behavior: E&B, 24*(3), 304–310. https://doi.org/10.1016/j.yebeh.2012.04.128.

Geertsema, E. E., Thijs, R. D., Gutter, T., Vledder, B., Arends, J. B., Leijten, F. S., Visser, G. H., Kalitzin, S. N. (2018). Automated video-based detection of nocturnal convulsive seizures in a residential care setting. *Epilepsia, 59*, 53–60. https://doi.org/10.1111/epi.14050.

Ghubaish, A., Salman, T., Zolanvari, M., Unal, D., Al-Ali, A., & Jain, R. (2021). Recent Advances in the Internet-of-Medical-Things (IoMT) Systems Security. *IEEE Internet of Things Journal, 8*(11), 8707–8718. https://doi.org/10.1109/jiot.2020.3045653.

Greene, B. R., Boylan, G. B., Reilly, R. B., de Chazal, P., & Connolly, S. (2007). Combination of EEG and ECG for improved automatic neonatal seizure detection. *Clinical Neurophysiology, 118*(6), 1348–1359. https://doi.org/10.1016/j.clinph.2007.02.015.

Gubbi, J., Kusmakar, S., Rao, A., Yan, B., T. O'Brien, M., & Palaniswami, M. (2015). Automatic detection and classification of convulsive psychogenic nonepileptic seizures using a wearable device. *IEEE Journal of Biomedical and Health Informatics*. https://doi.org/10.1109/jbhi.2015.2446539.

Halford, J. J., Sperling, M. R., Nair, D. R., Dlugos, D. J., Tatum, W. O., Harvey, J., French, J. A., Pollard, J. R., Faught, E., Noe, K. H., Henry, T. R., Jetter, G. M., Lie, O. V., Morgan, L. C., Girouard, M. R., Cardenas, D. P., Whitmire, L. E., & Cavazos, J. E. (2017). Detection of generalized tonic–clonic seizures using surface electromyographic monitoring. *Epilepsia, 58*(11), 1861–1869. https://doi.org/10.3410/f.731790939.793539134.

Harden, C., Tomson, T., Gloss, D., Buchhalter, J., Cross, J. H., Donner, E., French, J. A., Gil-Nagel, A., Hesdorffer, D. C., Smithson, W. H., Spitz, M. C., Walczak, T. S., Sander, J. W., & Ryvlin, P. (2017). Practice guideline summary: Sudden unexpected death in epilepsy incidence rates and risk factors. *Neurology, 88*(17), 1674–1680. https://doi.org/10.1212/wnl.0000000000004663.

Islam, M., Nooruddin, S., Karray, F., & Muhammad, G. (2022). Internet of Things: Device Capabilities, Architectures, Protocols, and Smart Applications in Healthcare Domain. *IEEE Internet of Things Journal*. https://doi.org/10.1109/jiot.2022.3228795.

Jallon, P. (2010). A Bayesian approach for epileptic seizures detection with 3D accelerometers sensors. 2010 *Annual International Conference of the IEEE Engineering in Medicine and Biology Society, EMBC'10*, 6325–6328. https://doi.org/10.1109/iembs.2010.5627636.

Jansen, K., Varon, C., Van Huffel, S., & Lagae, L. (2013a). Ictal and interictal respiratory changes in temporal lobe and absence epilepsy in childhood. *Epilepsy Research, 106*(3), 410–416. https://doi.org/10.1016/j.eplepsyres.2013.07.008.

Jansen, K., Varon, C., Van Huffel, S., & Lagae, L. (2013b). Peri-ictal ECG changes in childhood epilepsy: Implications for detection systems. *Epilepsy & Behavior, 29*(1), 72–76. https://doi.org/10.1016/j.yebeh.2013.06.030.

Jeppesen, J., Beniczky, S., Johansen, P., Sidenius, P., & Fuglsang-Frederiksen, A. (2015a). Detection of epileptic seizures with a modified heart rate variability algorithm based on Lorenz plot. *Seizure, 24*, 1–7. https://doi.org/10.1016/j.seizure.2014.11.004.

Jeppesen, J., Beniczky, S., Johansen, P., Sidenius, P., & Fuglsang-Frederiksen, A. (2015b). Exploring the capability of wireless near infrared spectroscopy as a portable seizure detection device for epilepsy patients. *Seizure, 26*, 43–48. https://doi.org/10.1016/j.seizure.2015.01.015.

Jeppesen, J., Fuglsang-Frederiksen, A., Johansen, P., Christensen, J., Wüstenhagen, S., Tankisi, H., Qerama, E., & Beniczky, S. (2020). Seizure detection using heart rate variability: A prospective validation study. *Epilepsia, 61*(S1), S41–S46. https://doi.org/10.1111/epi.16511.

Kamiŝalić, A., Fister, I., Turkanović, M., & Karakatiĉ, S. (2018). Sensors and Functionalities of Non-Invasive Wrist-Wearable Devices: A Review. *Sensors (Basel, Switzerland), 18*(6). https://doi.org/10.3390/s18061714.

Kumar, R., & Rajasekaran, M. P. (2016). An IoT based patient monitoring system using raspberry Pi. *2016 International Conference on Computing Technologies and Intelligent Data Engineering*. https://doi.org/10.1109/icctide.2016.7725378.

Kurada, A. V., Srinivasan, T., Hammond, S., Ulate-Campos, A., & Bidwell, J. (2019). Seizure detection devices for use in antiseizure medication clinical trials: A systematic review. *Seizure, 66*, 61–69. https://doi.org/10.1016/j.seizure.2019.02.007.

Li, W., Wang, G., Lei, X., Sheng, D., Yu, T., & Wang, G. (2022). Seizure detection based on wearable devices: A review of device, mechanism, and algorithm. *Acta Neurologica Scandinavica, 146*(6), 723–731. https://doi.org/10.1111/ane.13716/v1/decision1.

Lockman, J., Fisher, R. S., & Olson, D. M. (2011). Detection of seizure-like movements using a wrist accelerometer. *Epilepsy & Behavior, 20*(4), 638–641. https://doi.org/10.1016/j.yebeh.2011.01.019.

Looy, W. Van, Cuppens, K., Bonroy, B., & Vel, A. Van de. (2011). Detection of body movement using optical flow and clustering. In *Proceedings of International Conference on Positioning and Context-Awarenes*s (PoCA), Brussels, Belgium.

Masse, F., Bussel, M. Van, Serteyn, A., Arends, J., & Penders, J. (2013). Miniaturized wireless ECG monitor for real-time detection of epileptic seizures. *Transactions on Embedded Computing Systems, 12*(4). https://doi.org/10.1145/1921081.1921095.

Mathurin, G., Ntonfo, K., Ferrari, G., Raheli, R., & Pisani, F. (2012*). Low-complexity image processing for real-time detection of neonatal clonic seizures. IEEE Transactions on Information Technology in Biomedicine, 16*(3), 375. https://doi.org/10.1109/isabel.2010.5702898.

Meritam, P., Ryvlin, P., & Beniczky, S. (2018). User-based evaluation of applicability and usability of a wearable accelerometer device for detecting bilateral tonic-clonic seizures: A field study. *Epilepsia, 59*(Suppl 1), 48–52. https://doi.org/10.1111/epi.14051.

Milošević, M., Vel, A. Van de, Bonroy, B., Ceulemans, B., Lagae, L., Vanrumste, B., & Van Huffel, S. (2015). Automated detection of tonic–clonic seizures using 3-d accelerometry and surface electromyography in pediatric patients. *IEEE Journal of Biomedical and Health Informatics*. https://doi.org/10.1109/jbhi.2015.2462079.

Murthy, J. N., Van Jaarsveld, J., Fei, J., Pavlidis, I., Harrykissoon, R. I., Lucke, J. F., Faiz, S., & Castriotta, R. J. (2009). Thermal Infrared Imaging: A Novel Method to Monitor Airflow During Polysomnography. *Sleep, 32*(11), 1521–1527. https://doi.org/ 10.1093/sleep/32.11.1521.

Myers, L., Lancman, M., Laban-Grant, O., Lancman, M., & Jones, J. (2017). Socialization characteristics in persons with epilepsy. *Epilepsy & Behavior, 72*, 99–107. https://doi. org/10.1016/j.yebeh.2017.04.036.

Narechania, A. P., Garić, I. I., Sen-Gupta, I., Macken, M. P., Gerard, E. E., & Schuele, S. U. (2013). Assessment of a quasi-piezoelectric mattress monitor as a detection system for generalized convulsions. *Epilepsy & Behavior, 28*(2), 172–176. https://doi.org/10. 1016/j.yebeh.2013.04.017.

Osorio, I., & Manly, B. F. J. (2014). Is seizure detection based on EKG clinically relevant? *Clinical Neurophysiology, 125*(10), 1946–1951. https://doi.org/10.1016/j.clinph. 2014.01.026.

Osorio, I., & Manly, B. F. J. (2015). Probability of detection of clinical seizures using heart rate changes. *Seizure, 30*, 120–123. https://doi.org/10.1016/j.seizure.2015.06007.

Patterson, A. L., Mudigoudar, B., Fulton, S., McGregor, A., Van Poppel, K., Wheless, M. C., Brooks, L., & Wheless, J. W. (2015). SmartWatch by SmartMonitor: Assessment of Seizure Detection Efficacy for Various Seizure Types in Children, a Large Prospective Single-Center Study. *Pediatric Neurology, 53*(4), 309–311. https://doi. org/10.1016/j.pediatrneurol.2015.07.002.

Picard, R. W., Migliorini, M., Caborni, C., Onorati, F., Regalia, G., Friedman, D., & Devinsky, O. (2017). Wrist sensor reveals sympathetic hyperactivity and hypoventilation before probable SUDEP. *Neurology, 89*(6), 633–635. https://doi.org/ 10.1212/wnl.0000000000005286.

Poets, C. F. (2004). Apparent life-threatening events and sudden infant death on a monitor. *Paediatric Respiratory Reviews, 5*(SUPPL. A), S383–S386. https://doi.org/10.1016/ s1526-0542(04)90068-1.

Rémi, J., Silva Cunha, J. P., Vollmar, C., Bilgin Topçuoğlu, Ö., Meier, A., Ulowetz, S., Beleza, P., & Noachtar, S. (2011). Quantitative movement analysis differentiates focal seizures characterized by automatisms. *Epilepsy & Behavior, 20*(4), 642–647. https://doi.org/10.1016/j.yebeh.2011.01.020.

Renevey, P., Vetter, R., Krauss, J., Celka, P., & Depeursinge, Y. (2001). Wrist-located pulse detection using IR signals, activity and nonlinear artifact cancellation. *Annual Reports of the Research Reactor Institute, Kyoto University, 3*, 3030–3033. https://doi.org/10.1109/iembs.2001.1017437.

Ryvlin, P., Nashef, L., & Tomson, T. (2013). Prevention of sudden unexpected death in epilepsy: A realistic goal? *Epilepsia, 54*(s2), 23–28. https://doi.org/10.1111/epi. 12180.

Sabesan, S., & Sankar, R. (2015). Improving long-term management of epilepsy using a wearable multimodal seizure detection system. *Epilepsy & Behavior, 46*, 56–57. https://doi.org/10.1016/j.yebeh.2015.02.057.

Schulc, E., Unterberger, I., Saboor, S., Hilbe, J., Ertl, M., Ammenwerth, E., Trinka, E., & Them, C. (2011). Measurement and quantification of generalized tonic–clonic seizures in epilepsy patients by means of accelerometry—An explorative study. *Epilepsy Research, 95*(1–2), 173–183. https://doi.org/10.1016/j.eplepsyres.2011.02.010.

Security, A., Hireche, R., Mansouri, H., & Khan Pathan, A.-S. (2022). Security and Privacy Management in Internet of Medical Things (IoMT): A Synthesis. *Journal of Cybersecurity and Privacy, 2*(3), 640–661. https://doi.org/10.3390/jcp2030033.

Shankar, R., Jory, C., Tripp, M., Cox, D., & Hagenow, K. (2013). Monitoring nocturnal seizure in vulnerable patients. *Learning Disability Practice, 16*(9), 36–38. https://doi.org/10.7748/ldp2013.11.16.9.36.e1471.

Sheng, J., Liu, S., Qin, H., Li, B., & Zhang, X. (2018). Drug-Resistant Epilepsy and Surgery. *Current Neuropharmacology, 16*(1). https://doi.org/10.1016/j.eplepsyres.2003.10.001.

Simblett, S. K., Biondi, A., Bruno, E., Ballard, D., Stoneman, A., Lees, S., Richardson, M. P., & Wykes, T. (2020). Patients' experience of wearing multimodal sensor devices intended to detect epileptic seizures: A qualitative analysis. *Epilepsy & Behavior, 102*. https://doi.org/10.1016/j.yebeh.2019.106717.

Simon, R. P. (2010). Heart and lung in the postictal state. *Epilepsy & Behavior, 19*(2), 167–171. https://doi.org/10.1016/j.yebeh.2010.06.025.

Suzuki, S., Matsui, T., Kawahara, H., Ichiki, H., Shimizu, J., Kondo, Y., Gotoh, S., Yura, H., Takase, B., & Ishihara, M. (2009). A non-contact vital sign monitoring system for ambulances using dual-frequency microwave radars. *Medical and Biological Engineering and Computing, 47*(1), 101–105. https://doi.org/10.1007/s11517-008-0408-x.

Sveinsson, O., Andersson, T., Mattsson, P., Carlsson, S., & Tomson, T. (2020). Clinical risk factors in SUDEP: A nationwide population-based case-control study. *Neurology, 94*(4), E419–E429. https://doi.org/10.1212/wnl.0000000000009154.

Tamura, T., Maeda, Y., Sekine, M., & Yoshida, M. (2014). Wearable Photoplethysmographic Sensors—Past and Present. *Electronics, 3*(2), 282–302. https://doi.org/10.3390/electronics3020282.

Teohari, V., Ungureanu, C., Bui, V., Arends, J., & Aarts, R. (2014). Epilepsy seizure detection app for wearable technologies. *Conference Proceedings AmI, 2014*. https://doi.org/10.15760/honors.1523.

van Andel, J., Thijs, R. D., de Weerd, A., Arends, J., & Leijten, F. (2016). Non-EEG based ambulatory seizure detection designed for home use: What is available and how will it influence epilepsy care? *Epilepsy & Behavior, 57*(Pt A), 82–89. https://doi.org/10.1016/j.yebeh.2016.01.003.

van Andel, J., Ungureanu, C., Aarts, R., Leijten, F., & Arends, J. (2015). Using photoplethysmography in heart rate monitoring of patients with epilepsy. *Epilepsy & Behavior, 45*, 142–145. https://doi.org/10.1016/j.yebeh.2015.02.018.

Van de Vel, A., Cuppens, K., Bonroy, B., Milosevic, M., Van Huffel, S., Vanrumste, B., Lagae, L., & Ceulemans, B. (2013). Long-term home monitoring of hypermotor seizures by patient-worn accelerometers. *Epilepsy & Behavior, 26*(1), 118–125. https://doi.org/10.1016/j.yebeh.2012.10.006.

Van de Vel, A., Verhaert, K., & Ceulemans, B. (2014). Critical evaluation of four different seizure detection systems tested on one patient with focal and generalized tonic and clonic seizures. *Epilepsy & Behavior, 37*, 91–94. https://doi.org/10.1016/j.yebeh.2014.06.014.

van Elmpt, W. J. C., Nijsen, T. M. E., Griep, P. A. M., & Arends, J. B. A. M. (2006). A model of heart rate changes to detect seizures in severe epilepsy. *Seizure, 15*(6), 366–375. https://doi.org/10.1016/j.seizure.2006.03.005.

van Westrhenen, A., De Cooman, T., Lazeron, R. H. C., Van Huffel, S., & Thijs, R. D. (2019). Ictal autonomic changes as a tool for seizure detection: a systematic review. *Clinical Autonomic Research, 29*(2), 161–181. https://doi.org/10.1007/s10286-018-0568-1.

van Westrhenen, A., Souhoka, T., Ballieux, M. E., & Thijs, R. D. (2021). Seizure detection devices: Exploring caregivers' needs and wishes. *Epilepsy & Behavior, 116*, 107723. https://doi.org/10.1016/j.yebeh.2020.107723.

Verdru, J., & Van Paesschen, W. (2020). Wearable seizure detection devices in refractory epilepsy. *Acta Neurologica Belgica, 120*(6), 1271–1281. https://doi.org/10.1007/s13760-020-01417-z.

Vilella, L., Lacuey, N., Hampson, J. P., Rani, M. R. S., Sainju, R. K., Friedman, D., Nei, M., Strohl, K., Scott, C., Gehlbach, B. K., Zonjy, B., Hupp, N. J., Zaremba, A., Shafiabadi, N., Zhao, X., Reick-Mitrisin, V., Schuele, S., Ogren, J., Harper, R. M., & Lhatoo, S. D. (2019). Postconvulsive central apnea as a biomarker for sudden unexpected death in epilepsy (SUDEP). *Neurology, 92*(3), E171–E182. https://doi.org/10.1212/wnl.0000000000008586.

Wal, P., Wal, A., Verma, N., Karunakakaran, R., & Kapoor, A. (2022). Internet of medical things – the future of healthcare. *The Open Public Health Journal, 15*(1). https://doi.org/10.21275/sr23726161859.

Wu, S., Issa, N. P., Rose, S. L., Ali, A., & Tao, J. X. (2016). Impact of peri-ictal nurse interventions on postictal generalized EEG suppression in generalized convulsive seizures. *Epilepsy & Behavior, 58*, 22–25. https://doi.org/10.1016/j.yebeh.2016.02.025.

Xu, K., Lu, Y., & Takei, K. (2019). Multifunctional Skin-Inspired Flexible Sensor Systems for Wearable Electronics. *Advanced Materials Technologies, 4*(3), 1800628. https://doi.org/10.1002/admt.201800628.

Yamamoto, H., Okumura, A., & Fukuda, M. (2011). Epilepsies and epileptic syndromes starting in the neonatal period. *Brain and Development, 33*(3), 213–220. https://doi.org/10.1016/j.braindev.2010.10.009.

Chapter 4

The Improved Performance of a Double Pipe Heat Exchanger Using Different Configurations of Inserts in the Inner Pipe: A Review

Zainab M. Saleh[1,*]
Riyadh S. Al-Turaihi[1]
and Zena K. Kadhim[2]
[1]Department of Mechanical Engineering, University of Babylon, Babylon, Iraq
[2]Department of Mechanical Engineering, Wasit University, Wasit, Iraq

Abstract

Heat exchangers are crucial in engineering applications, and numerous researchers have been working to improve thermal performance and reduce dimensions and costs. This paper provides a comprehensive summary of various techniques for improving heat dissipation in double-pipe heat exchangers, making it a valuable resource for investigators, engineers, and professionals. The study evaluates the total heat dissipation coefficient, number of transfer units, exergy efficiency, pressure gradient, and thermal-hydraulic performance. The passive approach offers benefits such as no external power requirements and lower running costs compared to active alternatives. The results confirm significant advancements in heat transfer enhancement in double-pipe heat exchangers, with 47.8% of research focusing on passive techniques.

* Corresponding Author's Email: eng.zainab07@gmail.com

In: Advancements in Sustainability Systems II
Editors: H. H. Al-Kayiem, R. A. Jessam, Hasanain A. A. Wahhab et al.
ISBN: 979-8-89530-562-1

The study also examines the effectiveness of various insert devices, including a wire coil, helical coil, conical coil, helical fins, shell, and helical tube, and focuses on various cut shapes of twisted tape inserts in the double pipe heat exchanger. Recent studies show that inserts can increase heat transfer rates by altering their geometry or configurations, affecting pressure gradient and friction factors. Further theoretical study and numerical simulations are needed to further enhance heat transfer procedures.

Keywords: double pipe heat exchanger, heat transfer enhancement, twisted tape inserts, wire coil, conical ring, helical coil, helical tape

Abbreviation

PEF	Performance evaluation factor
PEC	Performance evaluation criteria
HNF	Hybrid nanofluid
TT	Twisted Tape
HTC	Heat Transfer Coefficients
PTTT	Perforated Triple Twisted Tape Insert
Re	Reynolds number
HTR	Heat Transfer Rate
Nu	Nusselt number
SCT	Spiral corrugated tube
CFD	Computational fluid dynamics
DPHE	Double pipe heat exchanger
PTT	Plain twisted tape
HE	Heat exchanger
LPM	Liter per minute
MFR	Mass flow rate
THP	Thermal and hydraulic performance
TR	Twisted ratio
De	Dean number

1. Introduction

Heat exchangers are comprehensively utilized in a wide range of industrial applications, including automotive, computer chips, refrigeration, and solar heat collector systems. These processes of heating and cooling expend a significant amount of energy, thereby contributing to the exacerbation of climate change. So, many countries have taken measures to promote global sustainability and inspire industries to participate in the Sustainable Energy Campaign. The optimization of heat transfer plays an important role in reducing energy wastage across different applications. The importance of energy conservation has led examiners to concentrate on heat transfer improvement, as an enhanced heat dissipation rate leads to augmented system performance, which is principal for thermal application systems (Mushatet et al. 2023). It is obtainable in multiple forms and is frequently utilized in engineering and industrial applications, encompassing factory settings, heating and cooling apparatus, and power plants, as well as different other industries like energy conservation, electricity, chemical, and petroleum sector, (Khafaji et al. 2022, and Ibrahim et al. 2022), as shown in Figure 1.

Circular tubes are frequently utilized in heat exchange equipment due to their ease of production and capability to resist high pressure. The research into dual tube heat devices expanded rapidly after 1928 and was divided into numerous categories. Some scholars are motivated by geometry change and the insertion of elements in both inner and outer tubes (Naphon et al. 2006; Córcoles et al. 2020). Other researchers explored numerous shapes of fins, and others explored several working fluids, such as nanofluids (Braga et al. 1999; Chun et al. 2008). Utilizing external forces such as magnetic field and vibration was the interest of some scholars in the subsequent years (Yildiz et al. 1995; Legay et al. 2012; and Kaska et al. 2018).

Prior research has focused on optimizing heat transmission in heat exchangers. Still, detailed inquiries into heat transfer in heat devices, specifically by employing different types of twisted tape, have been a shortage. Hence, the objective of the present investigation is to gather and categorize papers about different inserts and classify them based on the nature of the study of the shape of inserts and working fluid. This project aims to develop a strategic plan for investigating various cut shapes of twisted tape as an innovative method for heat transfer in heat exchangers. Hence, this research is a valuable resource for researchers, enabling them to explore enhanced heat transfer processes in heat exchangers using another shape of twisted tape (wavy edge) with perforated and different cut shapes to heat heavy fuel oil as best passive methodologies that are used to solve the problems in brick factories in Wasit city. These methodologies provide a means for researchers to comprehensively understand the various heat transfer directions within heat exchangers and different cut shapes of twisted tape modules.

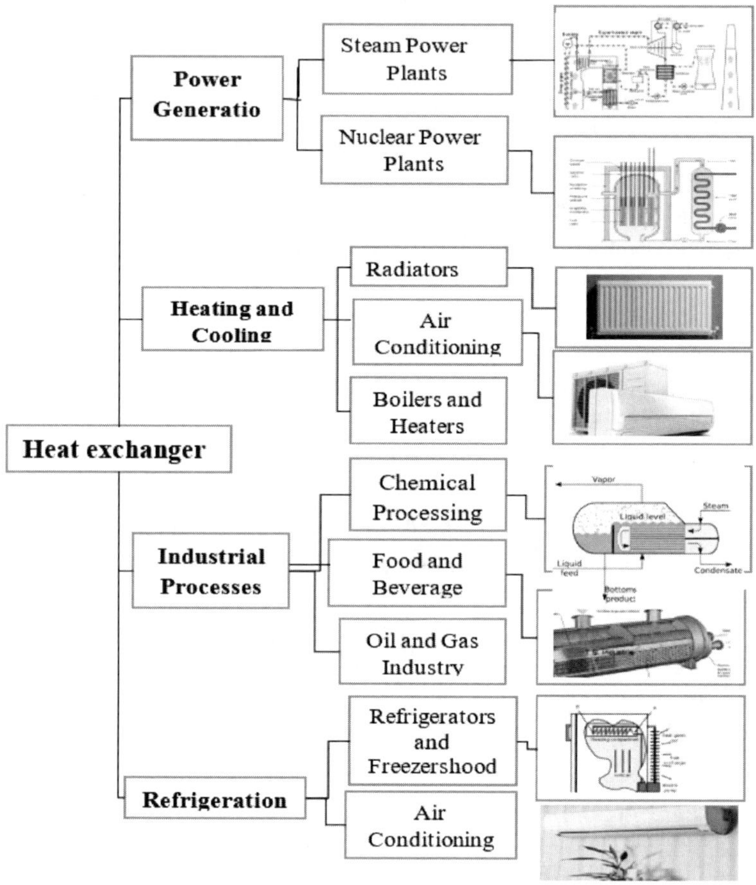

Source: Authors.

Figure 1. Some applications of heat exchangers.

2. Heat Exchanger Classification

Heat exchangers can be categorized in several ways based on their application, construction, or working principle, as shown in Figure 2. HE is important to transport heat between the cold and hot regions. Some types of HE (Kakaç et al. 2002, & Shah et al. 2003). The selection of heat exchangers utilized in a specific scenario will depend on numerous factors, such as the nature of the fluids involved, the desired heat transfer rate, and the spatial and resource constraints at hand.

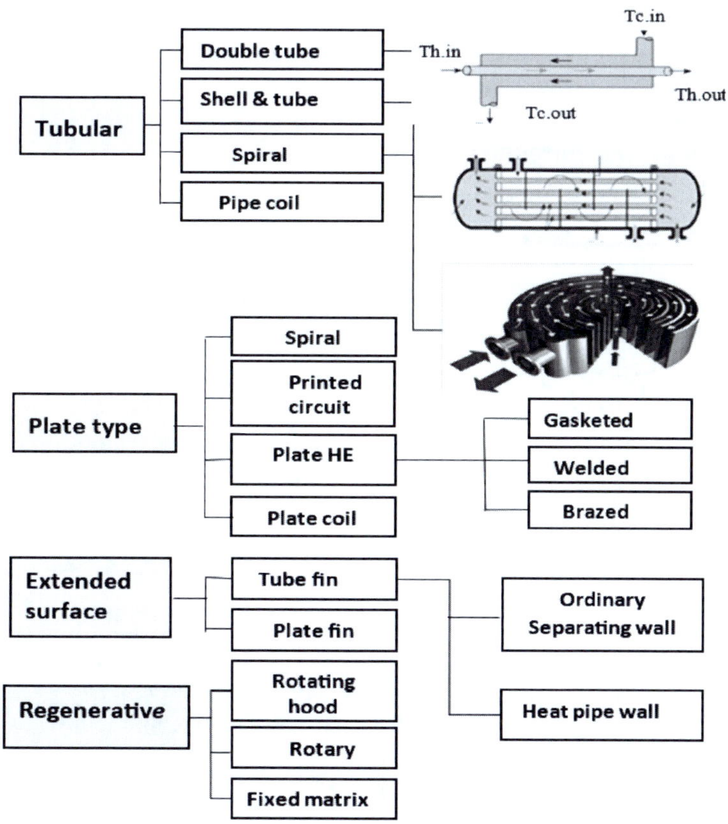

Source: Authors.

Figure 2. Classification of heat exchanger.

In terms of construction is divided into four subcategories: tubular, plate type, extended surface, and regenerative. The tubular heat exchangers section is divided into different types, which comprise shell and tube heat devices, spiral tubes, and pipe coil heat exchangers (Zohuri et al. 2017).

- Shell and tube heat devices. They are the most universally used type, comprising a cylindrical shell housing a bunch of tubes. While one fluid flows through the tubes, the other fluid circulates near the exterior of the tubes within the shell.
- Spiral heat devices: They are formed using two elongated, coiled metal strips, with one strip wound around the other to generate independent channels for fluid flow
- A pipe coil heat exchanger. They are a type of (HE) frequently used for transferring heat between two same or different fluids. It comprises a coil of pipes through which one fluid flows, while an alternative fluid surrounds the pipes and extracts or adds heat, and the fluids that are involved can be gases or liquids (Marzouk et al. 2023).

3. Heat Transfer Enhancement in Heat Exchanger

The methodologies utilized to enhance heat dissipation can be categorized into three principal classes, namely active, passive, and compound techniques (Al-Kayiem et al. 2017). Within the field of heat dissipation enhancement, there are three main techniques, as shown in Figure 3.

Types of heat transport enhancement for active, passive, and compound methods are shown in Figure 4.

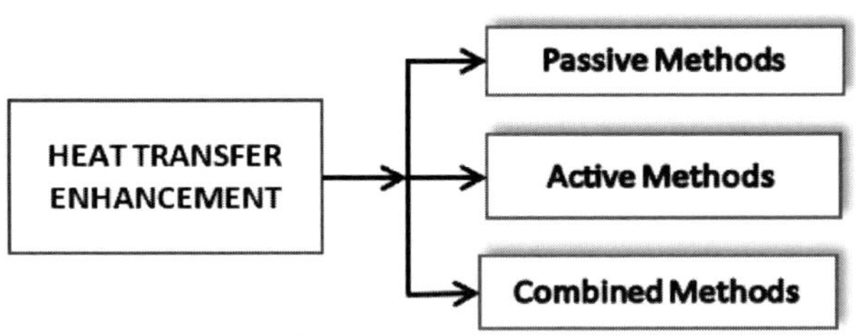

Source: Authors.

Figure 3. Techniques of heat transfer enhancement.

3.1. Actives Techniques

Their complexity exemplifies them in terms of both utilization and design. These techniques require the input of external power to achieve the desired variation of fluid flow and improvement in heat transfer rate. Due to the requirement of external power in different practical applications, their employment is limited (Nakate et al. 2015). In addition, fins or jet collisions of liquid and cooling gases can be found in order to cool bodies (Al-Maliki et al. 2021).

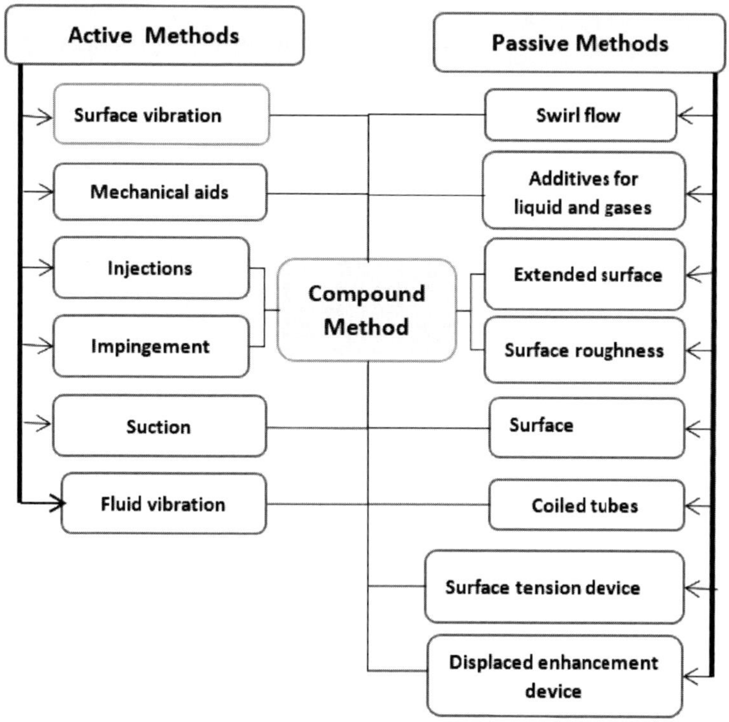

Source: Authors.

Figure 4. Types of heat transport enhancement for active, passive, and compound methods.

3.2. Passive Techniques

Concentrate on moderating the configuration of thermal systems to improve their thermal performance without the necessity for external energy sources (Rashidi et al. 2019). Different methods have been utilized to achieve this effect, including the employment of porous materials, inserts (such as turbulators, swirl flow tools, wire coils, and twisted strips), extended surfaces (such as fins), artificially roughened surfaces, protrusions and dimples, nanofluids, coil/helical/spiral tubes and displaced enhancement devices (Bovand et al. 2016, and Rashidi et al. 2017). The introduction of twisted tapes into the tubes of (HE) behaves as a passive technique to enhance heat transport by informing vortex generators, which interrupt the laminar sub-layer at the tube wall through reiterated changes in surface geometry. While this tape insertion technique increases heat transfer, it may also lead to a rise in the pressure drop (Kumar et al. 2017; Kumar et al. 2018; Kumar et al. 2017).

3.3. Compound Techniques

The simultaneous application of two or more of these techniques to accomplish greater enhancement in heat transport compared to individual usage. The incorporation of these techniques is termed a compound enhancement, as shown by Bergles et al. 2002, Bergles et al. 2011, and Rahul et al. 2017.

4. Heat Transfer Improvement by Inserts in Heat Exchanger

Several types of inserts frequently accomplish the enhancement and classification of heat exchangers (HE). For instance, twisted tape, helical screws, wire coils, baffles, ribs, plates, insert, convergent or divergent conical rings, mesh inserts, and conical rings are summarized in Table 1. Tube insert devices comprising twisted tape, extended surfaces, wire coils, and wire mesh inserts are considered the most crucial techniques of passive heat transfer augmentation methods. Among them, twisted tape with different cuts and wire helical tape is widely applied to others. This paper attempted to expose wire coils, Helical coil inserts with ribs, conical rings, helical fins, shells, and helical tubes, and focus on the important study of twisted tape (TT) inserts (Kiran et al. 2014).

Table 1. Main methods of the enhancement of heat exchangers.

Findings	Configuration	Method	Author
Higher turbulent heat dissipation coefficient, effectiveness, and friction factors compared to smooth tubes. A correlation was developed for calculating these coefficients.	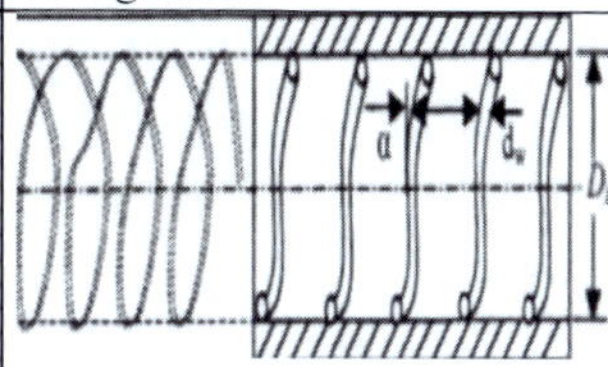	Wire coil	Abedin, M et al. (2018)
Helical ribs improved the Nusselt number by 12.45%, and the nanofluid concentration increased the friction factor.	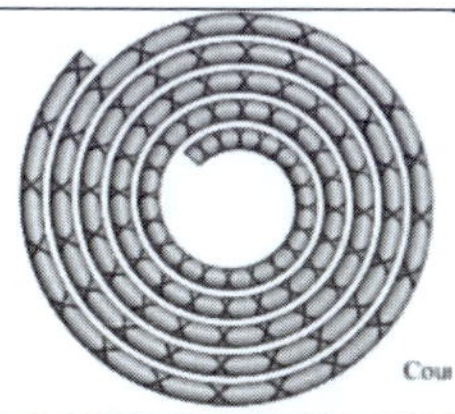	Helical coil insert with ribs.	Naphon, P et al. (2021)
conical rings in a heat exchanger tube, revealing a 278.2% improvement in heat transfer with 4 holes	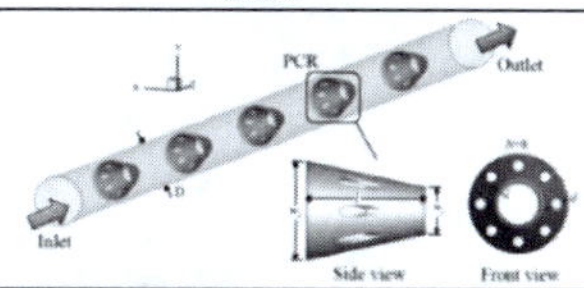	Conical Ring	Nakhchi M. E. et al. (2019)
Helical fins achieved better heat transfer rates and oil outlet temperatures, with an average improvement of 15% in heat transfer coefficient and 13% in effectiveness.	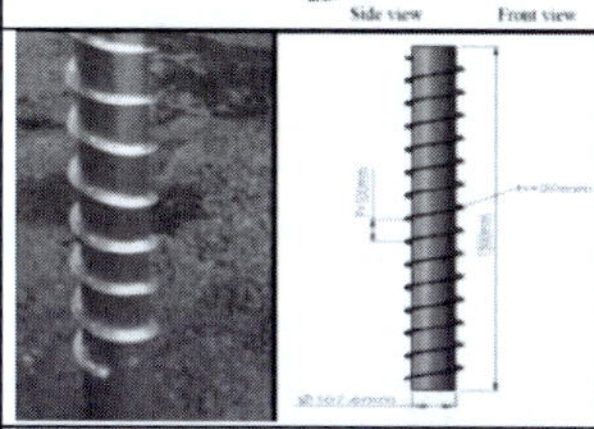	Helical Fins	Saleh, S et al. (2023)
Helical tube lengthening decreased heat transfer, oil outlet temperature effectiveness, and pressure drop	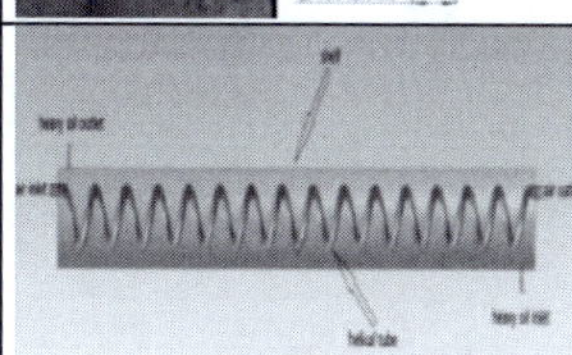	Shell and helical tubes	Saleh, S et al. (2023)
Nusselt number is found to be higher than plain heat exchangers, with the highest thermal efficiency factor being 1.63 times higher than simple heat exchangers.	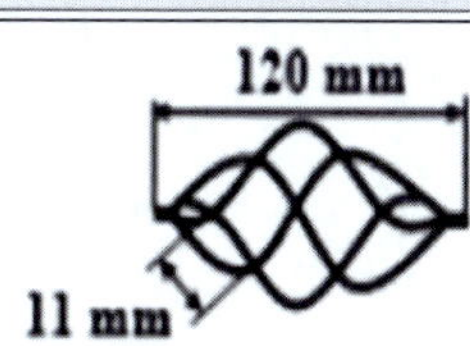	Twisted Tape	Khafaji et al. (2022)

Source: Authors.

4.1. Thermal Performance Factor

Performance factor (PF) and performance indicator a commonly utilized metrics to evaluate the effectiveness of several methods for enhancing heat transfer. It measures the ratio of improved heat dissipation to amplified skin friction coefficient at a specified Reynolds number (Re) and constant pumping power (Hwang et al. 2010).

$$\mathrm{PF} = \left(\frac{Nu_{tw}}{Nu_o}\right) / \left(\frac{f_{tw}}{f_o}\right)^{1/3} \quad (1)$$

where Nu_{tw}, Nu_{uo}, and f_{tw}, fo are the Nu and f for a tube with and without insert configuration, respectively, with calculated values for the Nu number, represented by Nu_{tw} and friction factor f, respectively:

$$\mathrm{Nu} = h.d/k \quad (2)$$

$$f = \Delta p/(\rho u^2)(L/d_H) \quad (3)$$

where Δp represents the pressure change across the test part, ρ represents the fluid density, L is the length of the tube, d_H is the hydraulic diameter of the tube, and u is the velocity of the fluid.

Another form of the performance factor, or performance indicator, is suggested by Al-Kayiem et al. 2011.

$$\mathrm{F} = \left(\frac{\mathrm{St_{tw}}}{\mathrm{St_o}}\right) / \left(\frac{\mathrm{f_{tw}}}{\mathrm{f_p}}\right)^{1/3} \quad (4)$$

4.2. Wire Coil Inserts

Wire-coil inserts have been utilized as an additional technique for augmenting passive heat transfer. These wire-coil inserts are normally used in various heat transfer equipment. They offer numerous advantages compared to other augmentation techniques, such as low cost in manufacturing, easy installation and removal, and the possibility of installation in present smooth tube heat exchangers. They also help mitigate fouling in chemical industries, refineries, and marine applications. Figure 5 illustrates a wire coil insert that can be utilized instead of a smooth one to realize higher heat transfers (Sonawane et al. 2016). (Abedin et al. 2018) explored the enhancement of turbulent heat dissipation flow through various tubes, including smooth and wire coil inserts, at low Reynolds numbers. Results showed a higher turbulent heat dissipation coefficient, effectiveness, and friction factors compared to smooth tubes. A correlation was developed for calculating these coefficients. Goudarzi et al. 2017 experimentally examined two wire coil inserts, and aluminum oxide nanofluid in ethylene glycol increased heat transfer rates by up to 9%. Coil inserts with varying nanofluid concentrations improved thermal performance by up to 5%. The study found no significant difference in thermal performance factors across Reynolds numbers. (Wang et al. 2020) studied water flow and heat dissipation by using a helical coil design with wire coil inserts and found that structural changes increased heat transfer but also increased pressure drop, indicating a direct impact on the thermal performance of the devices.

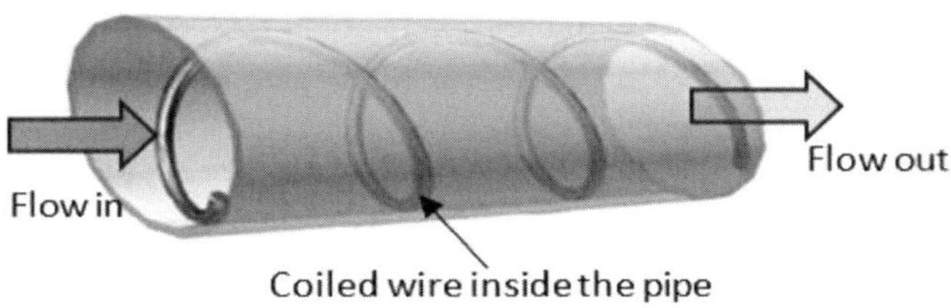

Source: Authors.

Figure 5. Wire coil insert.

4.3. Helical Coil Insert with Ribs

Zhang et al. 2020 calculated the flow and heat transfer in tubes under unequal heat flow conditions utilizing supercritical fluids, concentrating on the implementation of ribs. Their numerical analysis showed a significant improvement of 25.61% in heat transfer and 11.81% in the coefficient of friction. Virgilio et al. 2020 found that a discontinuous rib in a helical coil insert has a minor impact on pressure losses and turbulent kinetic energy generation. The discontinuous rib generates 50% and 35% lower turbulent kinetic energy and azimuthal velocity compared to a continuous arrangement. Additionally, the discontinuous obstacle decreases the skin friction coefficient and local heat dissipation enhancement. Naphon et al. 2021 showed numerical simulations on nanofluid flow and heat transfer in a helical coiled copper pipe with and without helical ribs, found centrifugal force promoted secondary flow, helical ribs improved Nusselt number by 12.45%, and nanofluid concentration increased friction factor. Bhattacharyya et al. 2021 investigated heat transfer and pressure drop characteristics in a heat exchanger tube with hybrid ribs using magnetic nanofluid (Fe3O4). It investigates the effects of different rib geometries on heat transfer and pressure drop for Reynolds numbers ranging from 3,000 to 22,000, finding excellent agreement between simulation results and established correlations.

Anuradha et al. 2020 numerically explored heat transport and friction coefficient in a dual tube counter flow heat device with two arrangements of inline triangular ribs on both the outer and inner pipes, as well as two arrangements of helical springs wrapped over the inner pipe. Al-Kayiem et al. 2017 explored the use of artificial roughening to improve the thermal performance of a double-pipe heat device. They suggested that ribs installed on the inner surface of the annulus ribs can enhance heat transfer but also increase pressure drop due to increased friction factor values. The results suggest that smaller pitches are more effective than larger ones, as shown in Figure 6. Al-Kayiem et al. 2011 developed an analytical procedure to study heat transport and pressure decline behavior in double-pipe heat exchangers. Results showed that ribs enhanced heat transfer by over 4 times, but increased pressure drop by 38 times. Annulus efficiency index and pitch distance were also investigated.

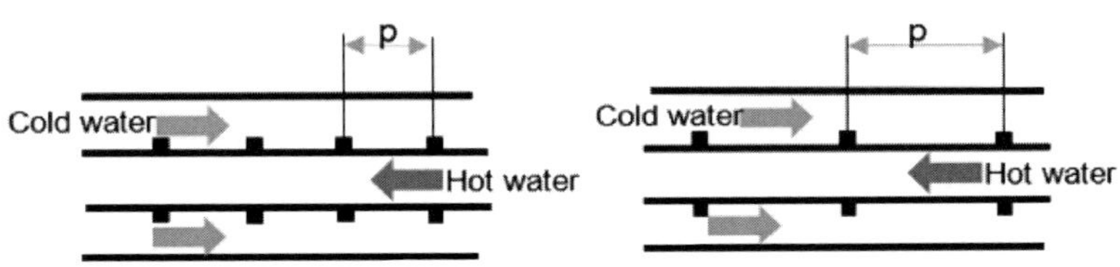

Source: Al-Kayiem et., 2017.

Figure 6. The ribbing cases.

4.4. Conical Ring

Passive heat augmentation devices like conical rings are widely used. Numerical investigations show that perforated and slotted inserts enhance the Nusselt number and friction factor. Solid ring inserts achieve the highest hydraulic performance (Banihashemi et al. 2022). Farajollahi et al. 2023 explored a two-phase turbulent nanofluid flow in a heat exchanger tube with a novel type of conical turbulator. The turbulators create vortices and recirculating currents, significantly affecting heat transfer. The number of turbulators increases the Nusselt number, but the presence of holes reduces the friction factor and pressure drop. The turbulators improve Performance Evaluation Criteria (PEC) by 43% compared to smooth tubes, with a maximum PEC of 1.43. Nakhchi et al. 2019 simulated 3D flow properties of Cu-water nanofluids using cribriform conical rings in a heat exchanger tube, revealing a 278.2% improvement in heat transfer with 4 holes.

4.5. Helical Fins

The study utilized a CFD model to compute outcomes for a helical fin with various fin pitches and heat exchanger effectiveness. They examined the impact of pipe surface roughness, fin pitches, and fin heights on heat transmission and pressure drops. Found that increasing surface roughness enhanced heat dissipation rate and pressure drops. Thermal performance improved by about 35% with helical fins. Reduced fin pitch led to more vortices, higher heat transfer rates, and pressure drops. Increasing fin height proved more effective than pitch alteration in increasing heat transfer rates (Mozafarie et al. 2019; Salman et al. 2021; and Mao et al. 2021). The study of the thermal-hydraulic performance of a DPHE was conducted using different pitch ratios of helical fins and Reynolds numbers. Outcomes revealed that helical fins increased Heat Transfer Rate (HTR) and pressure drops compared to smooth heat exchangers. As the pitch ratio decreased, Nu and friction factor improved by 156.6% and 54.6%, respectively. The thermal performance index improved with pitch ratio reduction, and OHTC increased by 35% (Sivalakshmi et al. 2021; Mule et al. 2017). Some others used different angles to study the effect of fin tip and fin root thickness on HE with helical fins. Results showed that OHTC increased by 53%, with a decrease above 20°. Nu increased by 3.29% at tip thicknesses 2 and 3.7 mm, and thermal performance increased by 2.72% (Zhang et al. 2023).

Saleh et al. 2023 examined the performance of a DPHE using helical fins to recover heat from exhaust emissions and warm heavy fuel oil in a brick plant. They found that increasing the oil mass flow rate increased heat dissipation rate, Nu (oil side), and overall heat transfer coefficient, while decreasing oil outlet temperature and heat transfer effectiveness. Heat exchangers with 20mm helical fins achieved better heat transfer rates and oil outlet temperatures, with an average improvement of 15% in heat transfer coefficient and 13% in effectiveness.

4.6. Shell and Helical Tubes

Surveyed SAHT with various cross sections using ANSYS FLUENT 2016 software with a high constant of the helical tube wall temperature, several coil diameters, and different grooving depths and found that decreasing coil diameter, increasing Nu and friction factor coefficient, and increasing Re and decreasing coil pitch improved OHTC development by about 45% compared to straight DPHE. HTR increased with fin height (Rezaei et al. 2019; Abu-Hamdeh et al. 2021; Najm et al. 2022; and Xu et al. 2022). In experimental studies, Kumar et al. 2020 and Dehghani et al. 2021 found that the largest pressure loss appears in helical microtubes with higher coil diameters, increasing volume flow rate in the coil side guided to decrease the effectiveness of HE. Heat transfer coefficients on the coil side were discovered to be higher for larger pitches compared to smaller pitches.

Numerically and experimentally studied the effect of two important parameters in helical HE: Dean number and torsion. When the Dean number, De, increased, the rate of heat transport increased by 75%, and Nu increased by 150%. HTR decreases with increased coil pitch (Krutova et al. 2020; Abdullah et al. 2023). Saleh et al. 2023 numerically analyzed a shell and helical tube heat exchanger to reduce fuel consumption in industrial furnaces. The hot gas was air, and the helical tube was constructed with various coil pitches and curvature ratios. The oil outlet temperature dropped with an increased oil mass flow rate. The study found that helical tube lengthening decreased heat transfer, oil outlet temperature effectiveness, and pressure drop.

Khata et al. 2021 studied laminar flow thermal and hydraulic efficiency of trapezoidal and straight counter heat exchangers with SiO2/water nanofluid and engine oil with Reynolds numbers from 1100 to 2300. Numerical simulation successfully modeled the smooth flow of nanofluids. Found that the Use of SiO2/water nanofluid significantly enhanced heat transfer. The numerical and experimental results were 0.0102% and 0.0208%. Also, the trapezoidal channel's influence on the Nusselt number was higher when water was used instead of nanofluid by Kadhim et al. 2020. Additionally, Kadhim et al. 2017 conducted a study on the augmentation of heat dissipation in (HE) using a transverse groove in the outer surface. The results showed that using corrugated tubes improved HTR by 58.24% to 59.55% at 40°C, 55.62% to 58.09% at 50°C, and 54.44% to 59.17% at 60°C. The experimental results showed a temperature difference improvement of 14.94% to 43.2% for both corrugated tubes.

4.7. Twisted Tape

The incorporation of twisted tapes (TT) within the exchanger tube serves to moderate the flow pattern and create turbulence, thereby disrupting the liquid layers and enhancing intermixing between them (Yakut et al. 2004). Kinds of twisted tapes are as follows: 1. Tapes with twist ratio, 2. Perforated Tapes, 3. Cut Tapes, 4. Doble twist tapes, 5. Dimple tapes, 6. Coupling, 7. V-cut tapes, 8. V-finned, 9. Tapered, 10. Clockwise twisted, 11. Nonuniform tapes, 12. Continuous spaced, 13. Trapaziodial, 14. Quadrapule cut, 15. U cut, 16. Multiple quadruple, 17. Double v-ribbed, and 18. Perforated and helical (Singh et al. 2019).

5. Heat Transfer Improvement by Twisted Tapes

Twisted tapes, shown in Figure 7, are used to augment heat dissipation in heat exchangers. The analysis examined the primary and operative mechanisms of heat dissipation augmentation, revealing a helical flow pattern in the tube, enhanced cutting speed and radial velocity, and increased mixing between the mainstream region and near-wall zone due to centrifugal force (Zhan, 2016).

Heat transfer enhancement is crucial for various applications, and twisted tapes (TT) are a notable device for this purpose. TT creates swirling liquid flow on the cylinder side, increasing velocity and mixing, resulting in a higher heat transfer coefficient. Proper clearance between TT and the tube wall is essential to prevent bypass flow and improve performance. Using TT can also create secondary flow patterns, reducing free flow area and viscosity effects (Dewan et al. 2004). TT is an instrumental tool used to augment the convective heat dissipation coefficient of a system. The authors aimed to increase convective heat dissipation by reducing the width of the boundary layer, resulting in the highest possible rate of heat dissipation. They suggested that reducing the boundary layer width to 30-35 units is the most effective method, ensuring optimal convective heat transfer and improving overall system efficiency (Bajestan et al. 2010). Some researchers found higher heat transfer coefficients and pressure drops in twisted-tube and straight-tube configurations with twisted tapes, with a decreased twist ratio increasing these parameters.

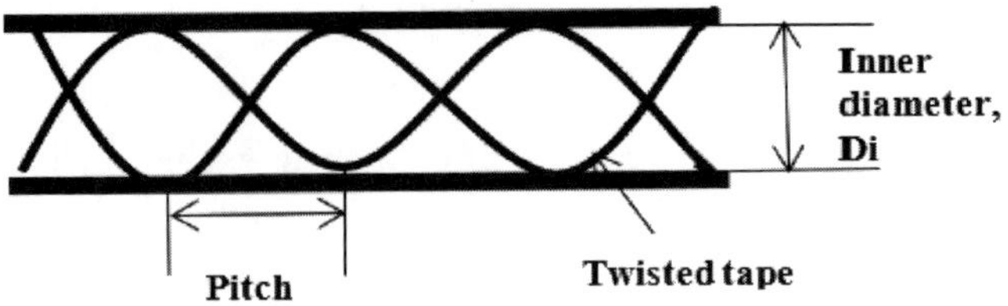

Source: Authors.

Figure 7. Structure sketch map of twisted tape.

Numerical investigations on twisted elliptical tube bundles showed higher heat transfer performance under crossflow conditions. This suggests that improving mechanisms for generating vortex and swirl patterns within pipes can disrupt boundary layer formation and convert laminar layers to turbulent ones (Farnam et al. 2018; Li et al. 2019; and Teng et al. 2010).

Hong et al. 2019 found that multiple twisted tapes to sinusoidal rib tubes (SRT) significantly improved heat dissipation from 27.4% to 39.5% but increased friction loss from 49.4% to 74.7%, suggesting SRT as a potential heat exchanger alternative. Piriyarungrod et al. 2018 examined the hydrothermal properties of heat exchanger tubes with multiple-twisted tapes (M-TT). Experimental results showed that the use of M-TT with various numbers of twists of 3, 4, 5, and 6 led to an improvement in heat transport and friction factor, compared to those noticed with a single twisted tape. Eiamsa-ard et al. 2009 used the finite volume method with the standard k-ε turbulence model and SST k-ω turbulence model to simulate and analyze the physical behavior of thermal and fluid flow in a tube fitted with loose-fit twisted tape under turbulent flow conditions. The outcomes showed a higher mean heat transfer rate of 73.6%.

Eiamsa-ard et al. 2010 experimentally examined thermal performance, friction factor, and heat transfer in tubes with twisted tape and wire coil pitch ratios. Results showed that compound devices, including TT and DI-coil, can improve thermal performance. Mogaji et al. 2013 conducted an experimental analysis of a horizontal tube containing a twisted tape in a two-phase flow of R134a. Results showed that it was appropriate when applied to the high vapor quality section of the evaporator and under mass velocities higher than 150 $kg/m^2{\cdot}s$. Dagdevir et al. 2021 experimentally explored the impact of perforated TT tabulators on thermal augmentation of the heat transfer rate and found that optimal thermal efficiency values were achieved with a pressure ratio of 0.25, η values of 1.42, 1.17, and 1.05, and volumetric ratios of 0:100, 20:80, and 40:60. (Wang et al. 2021) explored the flow and heat transfer (HT) features of

TT inserted into a tube. To augment the HT rate, they exploited silica-water nanofluids as the working fluid. The simulation results showed that the maximum augmentation in heat exchanger (HE) efficiency was realized with round and triangular tubes fitted with TT, with respective efficiency rises of 74.80% and 55.97%. This study also involved a fundamental representation of TT with perforation, as illustrated in Table 2. This demonstration aids in the comprehension of the design and functionality of the TT turbulator. Moya-Rico et al. 2020 evaluated the overall thermal efficiency of heat device systems by examining the influence of numerous TT configurations on the heat transport implementation, providing valuable insights into the system's performance.

Haddad et al. 2024 optimized flow and heat dissipation in solar collectors and heat exchangers using molten salts, using entropy generation minimization principles. It calculates the convective heat dissipation coefficient and pressure loss in tubes, determining the optimal Re, Nu, and tube diameter. (Bucak et al. 2022) examined the impact of turbulence (TT) on heat transfer in pipes using dimple-perforation patterns. It found that the highest dimpled and compressed protruded TT achieved optimal thermal performance factor values of 1.507 and 1.477 at a Reynolds number of 3000. Singh et al. 2020 investigated heat augmentation in a heat exchanger utilizing elliptical and circular inserted TT. The study focused on geometrical parameters like height ratio and Reynolds numbers. The optimum thermal performance factor (TPF) was achieved with a height ratio of 0.45 and a Re of 3000, resulting in a TPF value of 2.43.

Chougule et al. 2015 also explored the use of nanofluids in transition flow by applying helical twisted tape inserts. Nalavade et al. 2022 evaluated the impact of heat transfer and friction factor characteristics on nonconductive TT inserts in turbulent flow conditions with Re from 6000 to 22,000, achieving the highest improvement efficiency of 2.18 at a Reynolds number of 7000 and a pitch ratio of 1.63. Waghole et al. 2016 showed research on the utilization of TT with silver nanofluids. The investigation aimed to examine the impact of using silver nanofluids as the main fluid in heat exchangers equipped with TT. The outcomes of this study provide an understanding of the potential benefits of exploiting silver nanofluids in heat transfer applications.

Table 2. Summary of recent research on twisted tape with different cutting forms

Author (s)	Investigation Method	Working fluid	Main parameters examined	Main conclusion
V-cut configuration of twisted tape 				
Dandoutiyaet al. 2021	Numerical	water	Nusselt number, friction factor thermal performance factor (TPF)	utilizing v-cut TT improved fluid mixing and increased the Nu by 2.45 times and (f) by 9.22 times compared to plain tubes, with the highest (TPF) at Re = 5500
Souayeh et al. 2021	Numerical	Air	Solar air heater tube thermal energy transport coefficient twisted ratio (TR)	Long TT has a higher thermal energy transport coefficient compared to short-length tapes. The twist ratio has a greater influence on the Nu number than the angular cut angle.
Hamza et al. 2023	experimental	Hybrid nanofluid	tube heat exchanger Nusselt number,	led to a maximum increase in Nu number of 47.17%, a 29% enhancement in heat transfer, and a maximum thermal performance
Singh et al. 2023	Numerical	Fluid	heat exchanger, HTR, and friction factor, v-cut angles of 15°, 30°, and 45°	The optimized design demonstrates the effectiveness of the heat exchanger optimization.
Elmasry et al. 2023	Numerical	Fluid	Photovoltaic thermal (PVT) system pitch TT	The highest overall thermal and exergy efficiencies of 75.46% and 16.34% were obtained for the PVT system pitch of 100 mm and at Re = 2000
Sun et al. 2023	Numerical	Fluid	different numbers and turbulator pitches	Increasing turbulator pitch decreased (*h*) by 30.21-19.32% near the entrance, middle zone, while increasing it by 5.52-0.26% for the outlet zone of the tube.
Square and rectangular cut configuration of twisted tape				
Wijayanta et al. 2019	Numerical	Fluid	Various pitch ratios	thermal performance, with the highest increases reaching 81% and 3.3 times greater than a smooth tube thermal performance factor of 1.23-
Perng et al. 2022	Numerical	Fluid	circular tube	The simulations showed that the placement of a twisted tape with square cuts can effectively promote thermal convection in the tube,

Author (s)	Investigation Method	Working fluid	Main parameters examined	Main conclusion
Kadbhane et al. 2023	Numerical	Fluid	various pitch ratios, Re, depth, and width ratios	Heat transfer increases with a decreasing twist ratio At (TR) of 2, showing the highest Nu and *f*.
Nakhchi et al. 2019	Numerical	Fluid	rectangular cut twisted tape (RCT) friction factor f Heat transfer coefficient (HTC) Thermal performance (TP)	*f* enhances up to 34.8% in comparison with conventional twisted tapes. (HTC) increases up to 33.26% with increasing b/w from 0.25 to 0.75. (TP) is highest at b/w = 0.75, c/w = 0.5.
Barzegar and Vahid 2019	Numerical	Fluid	rectangular cut twisted tape (RCT)in (DPHE) constant TR, different cut depth and width ratios	The ultimate thermal performance factor of 1.46 is realized with RCTs.
Nakhchi et al. 2020	experimental		heat transfer and friction factors	Increasing the cut ratio from 0.25 to 0.90 enhances the Nusselt number by 177.4%, and thermal performance varies from 1.63 to 1.44.
Rahman et al. 2023	Experimental		Reynolds number of 12,500	TTIs with high porosity consistently reduced temperature differences compared to conventional TTIs. At Re equal 12,500, exhibited a significant increase in turbulent kinetic energy of about 53.1%.
Hong et al. 2023	Numerical	liquid-gas	Space ratios (S/D), longitudinal perforation length ratios (L/D)	Both Nu and f increased with an increase in the number of TTs and decreased with increasing S/D and L/D. Enhancement of η by approximately 7.9%.
Singh et al. 2022	Numerical		Relationship between heat dissipation and friction factor (f) using V-shaped TT.	The addition of swirl flow generated an 84% higher heat transfer rate (HTR) compared to plain tubes, providing a comprehensive understanding of heat transport characteristics with V-shaped twisted tape inserts.
Wijayanta et al. 2019	Numerical		Thermal hydraulic performance	The results show that the STT increases the Nusselt number and friction factor by 45.4-80.7% and 1.23 times, respectively, compared to the plain tube.

Table 2. (Continued)

Author (s)	Investigation Method	Working fluid	Main parameters examined	Main conclusion
Circle and semicircular cut configuration of twisted tape				
Sivakumar et al. 2020	Numerical		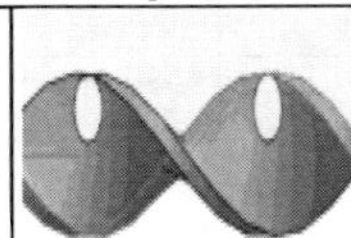	The CFD analysis shows that the heat transfer rate is enhanced by 1.1-1.3 times compared to basic twisted tape and circular cut twisted tape.
Das et al. 2022	Numerical		Twist ratio and cut diameter 	Smaller twist ratios increased heat transfer due to extra swirl, and larger cut diameters of about 11 mm resulted in higher Nusselt numbers.
Diwaker et al. 2023	Numerical		Different flow rates. 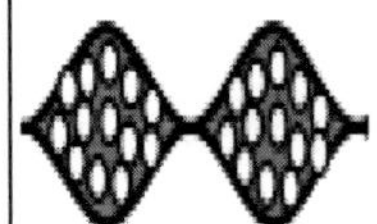	The use of circular and semicircular cuts led to significant improvements in thermohydraulic performance metrics.
Elliptical cut configuration of twisted tape				
Ganorkar et al. 2015	Experimental		Fixed twist ratio, Re (5000 to 22000)	The average Nu and f within the tube, when equipped with an elliptical-cut twisted tape, exhibit an increase as the major-to-minor axis ratio decreases.
Ahmadi et al. 2020	Numerical	nanofluid	ratio of short diameter to twisted tape width and the ratio of long diameter to short diameter	The highest HTP was achieved with a concentration of 4% Cupric Oxide nanoparticles higher thermal efficiency factor21%compared to water
Khfagi et al. 2022	Numerical	nanofluid	-----	It appeared that heat transport can be enhanced by using TT inserts and increasing the mass concentration of nanoparticles.

Author (s)	Investigation Method	Working fluid	Main parameters examined	Main conclusion
Trapezoidal cut configuration of twisted tape				
Thote et al. 2023	Numerical	Fluid	Reynolds Numbers 830 to 1990, different twist ratios	The presence of TT leads to significantly higher values of the 'frictional factor.' Tapes with the lowest twist ratio. Exhibit greater effectiveness
Thote et al. 2023	Numerical		Re = 200 to 2000, Nusselt Number	Nu experiences a gradual increase as the tape twist ratio decreases. Heat transfer efficiency can be enhanced by approximately 69%.
Thote et al. 2024	------	------	-------	HT enhancement techniques help save energy but can increase friction losses and pressure drop.
Several kinds of cut configurations of twisted tape				
Zheng et al. 2021	Experimental	Fluid	Reynolds number of 1800.	The performance evaluation criterion (PEC) reached a maximum value of 1.54
Diwaker et al. 2023	Numerical	Fluid	(Triangular and semicircular cuts) with $x/b = 4.5$, $\varphi = 8$ mm, and $B/Y = 1.1$.	Significant enhancement in thermo-hydraulic performance metrics, including the Nu and f. The highest thermal performance factor was 1.57.

Source: Authors.

Kamel et al. 2018 analyzed a DPHE with and without a full-length tight-fit twisted tape insert. Results showed that the twisted tape with a TR = 0.5 provides the greatest benefit, with maximum enhancement percentages of 55.4%, 203.2%, 183.5%, and 28.8%. Pressure losses increase with smaller twist ratios, peaking at 5.826 kPa for twisted tape with TR 0.5. Kadbhane et al. 2024 numerically investigated the heat dissipation performance of a horizontal heat tube with hexagonal perforated TT inserts. The optimized solutions for Nu, friction factor, and thermal performance factor are obtained using response surface methodology, with the best results achieved with perforated twisted tape inserts. Salim 2024 examined twisted double-tube heat exchangers, which showed enhanced heat transfer efficiency, reduced pressure drops, and smaller equipment size, with a decrease in twisting ratio resulting in increased Nu and heat transfer coefficient.

On the compound nanofluids and twisted tapes, Al-Kayiem et al. 2020 investigated heat and fluid flow characteristics in pipes induced with multiple plain twisted tape inserts utilizing water and 0.1 vol.% TiO2/water nanofluid. The results showed that tubes fitted with quintuple twisted tape inserts performed the best in augmenting heat transfer, with Nusselt numbers up to 110.8% for nanofluid flow and 100.0% for water flow. As the number of twisted tapes increased, the Nusselt number also increased. The pressure drops and the friction factor increase significantly with twisted tape inserts and slightly with nanofluid. The compound technique of twisted tape inserts and nanofluid is recommended for heat exchanger application due to the significant enhancement in heat transfer. Mushatet et al. 2018 examined the impact of twin twisted tape and nanofluid in a heated circular tube to reduce the size of heat transfer equipment and energy costs using Al_2O_3-water nanofluid. The study found that a 1.0 twist ratio and 60-degree cut angle led to the most significant increase in heat transfer efficiency. Table 2 summarizes the research that dealt with twisted tapes as a passive heat transfer method.

6. Effect of Twisted Tape Design Parameters

6.1. The Impact of Twisting Ratio Y/w and Twisting Angle

Researchers have investigated the influences of twisting ratio and twisting angle on heat transfer (HT) and pressure losses in tape eddy flows. These findings aim to identify the optimal design of tapes that reduce friction losses in laminar and turbulent flows (Ajarostaghi et al. 2022; Chavan et al. 2017; and Ponnada et al. 2019). Additional experiments have observed the influence of twisting tape inserts on heat exchangers in laminar tube flow with various pitches and twisting ratios. Compared to plain tubes, the tube with twisted tape inserts showed an increase in heat dissipation rate at different twisting ratios (Jassim et al. 2017). Additionally, concentric tube heat exchangers with twisting tapes demonstrated an increase in the Nu number due to the amplified swirl degree of flow. Copper pipes with tape inserts and a square-edged intake were also experimented by Meyer et al. 2017.

6.2. The Impact of Tape Surface Shape

Sivashanmugam et al. 2007 examined various helical screw inserts and full-length helical screw segments with varying twisting ratios. Results showed that right-left helical screw inserts increased the heat transmission rate more than straight helical twists, with maximum performance ratios of 2.85 and 2.97. Ananth et al. 2013, Gawande et al. 2017 Hajare et al. 2016 and Chavan et al. 2017 examined twisted tape inserts with variation-shaped holes, including square, elliptical, and rectangular holes, and twisting tape (TT) with a rod and spacer. Hasanpour et al. 2016 observed that twisting tapes with cuts influence cross-hollow twisting tape and multi-twisted tapes on the augmentation of thermohydraulic efficiency. Several investigators, such as Abeens et al. 2020, and Salman et al. 2013, examined twisted tape inserts with strips and were baffled.

6.3. The Impact of Tape Length and Number

Several researchers have conducted different studies to examine the influence of twisted tape length on performance. Yadav 2009 examined the impact of half-length TT inserts on pressure loss and heat exchange in double-pipe U-bend heat exchangers. Comparing these inserts to traditional HE, we found that heat transfer rates amplified approximately 40%. Smooth tubes revealed better performance than half-length twisted tapes in terms of pressure drop. Standard heat exchangers exhibited a thermal superiority of 1.3 to 1.5 times over half-length twisting tape. Eiamsa-ard et al. 2012 analytically and empirically examined a fading swirl turbulent flow created by short, twisted tapes positioned at the entrance of the test section with three separate twisting ratios (3, 4, and 5). The results exhibited that short tape inserts had inferior thermal performance in the interior in the Reynolds range of 5200-15,300 and at similar twist ratios as full-length tape.

Gugulothu 2020 utilized computational fluid dynamics and ANSYS Fluent 17.1 to investigate a ribbed tube with numerous twisted tape inserts under uniform heat flux in a turbulent region. The overall performance of full-length twisted tape was established to be higher than the half-period tape in terms of heat transfer (HT), friction factor, and PF. Based on their findings. Zhu et al. 2015 testified that triple-twisting tape inserts can raise resistance by 8.7-6 times when compared to single tape inserts while also augmenting heat transfer ability by 1.8-4.5 times. Promthaisong et al. 2017 exhibited a study on the thermal performance of spiraling corrugated tubes with five-channel twisting tape. Equations were employed to analyze the influence of dual twisted tape inserts with various pitches on turbulent heat dissipation and pressure loss. Aghaie et al. 2019 applied ANSYS Fluent (16.1) to survey heat dissipation inside a circular pipe with nanofluid under constant heat flux. The outcomes showed that the relative Nusselt number reduces as the Reynolds number rises for all analyzed scenarios.

7. Compound Nanofluids and Twisted Tape in Enhancing Heat Transfer

Sarafraz et al. 2016 employed a biological nanofluid to enhance forced convection in a dual-tube heat device. A mixture of ethylene glycol and water in a 50:50 volume ratio was utilized as the base fluid. The volume fraction of the nanomaterial was 0.1%, 0.5%, and 1%. The results indicated that when the volume percentage was raised to 1%, there was a significant 67% rise in the heat transfer coefficient. Wang et al. 2021 used silica-water nanofluids to improve heat transfer efficiency in heat exchangers. Numerical simulations showed that triangular tubes had better flow resistance, heat transfer impact, and comprehensive performance. Open circular holes on twisted tape improved the comprehensive evaluation index. Porous twisted tape, nanofluids, and triangular tubes improved system economy and heat exchange ability. Prasad et al. 2015 utilized Al2O3 nanofluid in combination with helical tape inserts. The researchers determined that the use of helical tape inserts with nanofluid resulted in improved performance.

Arani et al. 2013 conducted studies to examine the impact of TiO2 nanoparticle size on heat transmission properties. Their results demonstrated that nanoparticles with a diameter of 20 mm exhibited the highest TPF within the investigated range of Reynolds numbers and volume percentages of nanoparticles. Hormozi et al. 2016 examined the impact of surfactants on the thermal performance of hybrid nanofluid. The utilization of Alumina-Silver nanofluid with Sodium Delecyl Sulfate (SDS) anionic surfactant resulted in a 16% increase in Thermal Performance Factor compared to that of pure distilled water. Sundar et al. 2010 Determined the thermophysical properties of A12O3 nanofluid through experiments at different volume concentrations and temperatures. Convective heat transfer coefficient and friction factor data are also determined. The results show that the heat transfer coefficient and friction factor of 0.5% volume concentration are 33.51% and 1.096 times higher than water flow.

Conclusion

A review of the research on the numerous kinds of inserts under diverse settings and configurations has been examined. The following is a concise summary of the key conclusion.

1. To improve the turbulence of the heat exchanger's thermal performance and heat transfer characteristics, intensity, and the fluid's swirl flow near the wall surface, the twisted tape is regarded as an important variable.
2. The crucial requirement for improving overall performance is effective geometry, which has the power to intensify the turbulence produced by different insert devices like twisted tapes.
3. The twist ratio of the twisted tape is important for refining local heat transmission over the length ratio; a lower twist ratio is often associated with superior fluid mixing in addition to enhanced output. Therefore, twisted tape with a helical tube at a frictional pressure drop of 0.35 can increase the maximum equivalent heat transfer rate by up to 400% because of the swirl flow of fluid along the wall surface and the degree of turbulence in comparison to smooth tubes.
4. The wire coil inserts provide several advantages over other enhancement methods. For equivalent Reynolds numbers, the wire coil-inserted tubes' heat transfer efficacy can improve by up to 1.25

times compared to the smooth pipe. Nevertheless, when coil inserts with nanofluid at concentrations of 0.08%, 0.5%, and 1% are used instead of coil inserts alone, the thermal performance is improved by up to 5%.

5. Better thermal performance was also achieved by structurally altering the standard helical coil with a wire coil insert arrangement, even though the pressure drop rose by up to 115% as the curvature ratio increased. According to published research, wire coil inserts can increase heat transmission by up to 9% at a friction factor of 11.18%.
6. The greatest heat transfer boost is seen in twisted tape inserts, which have a 400% enhancement, followed by a wire coil at 9%.

Prior research has mainly concentrated on examining the enhancement of dual tube heat devices by using different shapes of cut twisted tape. Disregarding the possible benefits of using another shape of twisted tape and other working fluids as heavy oil, as an example. Hence, this pioneering research project suggests that future investigations might focus on a more detailed analysis of cutting shape twisting tape, enabling possible enhancements regarding heat transport performance to heat heavy oil in double pipe counter flow heat exchange.

References

Abdullah, M. S., & Hussein, A. M. (2023). Impact of coil pitch on heat transfer enhancement of a turbulent flow of α-Al2O3/DW nanofluid through helical coils. *Thermal Science*, (00), 131-131. https://doi.org/10.2298/tsci230227131a.

Abedin, M. Z., & Sarkar, M. R. (2018). Experimental study of heat transfer enhancement through a tube with wire-coil inserts at low turbulent Reynolds number. *International Journal of Engineering Materials and Manufacture*, *3*(3), 122-133. https://doi.org/10.26776/ijemm.03.03.2018.01.

Abeens, M., Meikandan, M., Sheriff, J., & Murunganadhan, R. (2020). Experimental analysis of convective heat transfer on tubes using twisted tape inserts, louvered strip inserts, and surface-treated tubes. *International Journal of Ambient Energy*, *41*(5), 540-546. https://doi.org/10.1080/01430750.2018.147626.

Abu-Hamdeh, N. H., Alsulami, R. A., Rawa, M. J., Aljinaidi, A. A., Alazwari, M. A., Eltaher, M. A., ... & Safaei, M. R. (2021). A detailed hydrothermal investigation of a helical micro double-tube heat exchanger for a wide range of helix pitch lengths. *Case Studies in Thermal Engineering*, *28*, 101413. . https://doi.org/10.1016/j.csite.2021.101413.

Aghaie, A., & Rabienataj Darzi, A. A. (2019). Heat transfer and pressure drop of Al2O3 /water nanofluid in a tube equipped with double twisted tape inserts with different pitch ratios. *Heat Transfer—Asian Research*, *48*(1), 233-253. https://doi.org/10.1002/htj.21380.

Ahmadi, K., Khanmohammadi, S., Khanmohammadi, S., Bahiraei, M., & Bach, Q. V. (2020). Heat transfer assessment of turbulent nanofluid flow in a circular pipe fitted with elliptical-cut twisted tape inserts. *Journal of Thermal Analysis and Calorimetry*, 1-14. https://link.springer.com/article/10.1007/s10973-020-10338-1#citeas.

Ajarostaghi, S. S. M., Zaboli, M., Javadi, H., Badenes, B., & Urchueguia, J.F. (2022). A review of recent passive heat transfer enhancement methods. *Energies*, *15*(3), 986. https://doi.org/10.3390/en15030986.

Al-Kayiem, H. H., & El-Rahman, M. (2011). Ribbed double pipe heat exchanger: analytical analysis. *Journal of Engineering Science and Technology*, *6*(1), 39-49. https://www.researchgate.net/profile/hussain-al-kayiem/publication/49941764_

Al-Kayiem, H. H., Ekhwan, A. B., & Muhi, L. N. (2017). Augmentation of ribs turbulators height on the hydrothermal performance of double pipe heat exchanger. *Journal of Engineering Science and Technology*, *12*(2), 548-563. https://www.researchgate.net/profile/hussain-al-kayiem/publication/316507770.

Al-Kayiem, H. H., Kassim, M. S., & Taher, S. T. (2020). Applications of compound nanotechnology and twisted inserts for enhanced heat transfer. *Inverse Heat Conduction and Heat Exchangers*, 23. https://doi.org/10.5772/intechopen.93359.

Al-Maliki, W. A. K., Al-Hasnawi, A. G. T., Abdul Wahhab, H. A., Alobaid, F., & Epple, B. (2021). A comparison study on the improved operation strategy for a parabolic trough solar power plant in Spain. *Applied Sciences*, *11*(20), 9576.

Ananth, J., & Jaisankar, S. (2013). Experimental studies on heat transfer and friction factor characteristics of thermosyphon solar water heating system fitted with regularly spaced twisted tape with rod and spacer. *Energy Conversion and Management*, *73*, 207-213. https://doi.org/10.1016/j.enconman.2013.04.022.

Arani, A. A., & Amani, J. J. E. T. (2013). Experimental investigation of diameter effect on heat transfer performance and pressure drop of TiO2–water nanofluid. *Experimental thermal and fluid science*, *44*, 520-533. https://doi.org/10.1016/j.expthermflusci.2012.08.014.

Anuradha, P., Ahmad, M., & Gupta, G. (2020). Numerical Investigation of Heat Transfer Enhancement in Double Pipe Counter-flow Heat Exchanger Fitted with Triangular Ribs and Spring as Inserts. *IUP Journal of Mechanical Engineering*, *13*(1). https://openurl.ebsco.com/results?sid=ebsco:ocu:record&bquery=is+0974-6536+and+vi+13+and+ip+1+and+dt+2020.

Banihashemi, S., Assari, M. R., Javadi, S. M., & Vahidifar, S. (2022). Thermal performance investigation in circular tube with stationary and rotating conical-obstacle inserts. *Journal of Thermophysics and Heat Transfer*, *36*(2), 242-255. https://doi.org/10.2514/1.t6199.

Barzegar, A., & Jalali Vahid, D. (2019). Numerical study on heat transfer enhancement and flow characteristics of double pipe heat exchanger fitted with rectangular cut twisted tape. *Heat and Mass Transfer*, *55*(12), 3455-3472. https://link.springer.com/article/10.1007/s00231-019-02667-1#citeas.

Bergles, A. E. (2002). ExHFT for fourth generation heat transfer technology. *Experimental Thermal and Fluid Science*, *26*(2-4), 335-344.

Bergles, A. E. (2011). Recent developments in enhanced heat transfer. *Heat and mass transfer*, *47*(8), 1001-1008. https://doi.org/10.1016/s0894-1777(02)00145-0.

Bhattacharyya, S., Vishwakarma, D. K., Roy, S., Dey, K., Benim, A. C., Bennacer, R., ... & Huan, Z. (2021). Computational investigation on heat transfer augmentation of a circular tube with novel hybrid ribs. In *E3S Web of Conferences* (Vol. 321, p. 04012). EDP Sciences. Https://doi.org/10.1051/e3sconf/202132104012.

Bajestan, E. B., Niazmand, H., Duangthongsuk, W., & Wongwises, S. (2011). Numerical investigation of effective parameters in convective heat transfer of nanofluids flowing under a laminar flow regime. *International journal of heat and mass transfer*, *54*(19-20), 4376-4388. https://doi.org/10.1016/j.ijheatmasstransfer.2011.05.006.

Bovand, M., Rashidi, S., & Esfahani, J. A. (2016). Heat transfer enhancement and pressure drop penalty in porous solar heaters: numerical simulations. *Solar Energy*, *123*, 145-159.. https://doi.org/10.1016/j.solener.2015.10.054.

Braga, C. V. M., & Saboya, F. E. M. (1999). Turbulent heat transfer, pressure drop and fin efficiency in annular regions with continuous longitudinal rectangular fins. *Experimental thermal and fluid science*, *20*(2), 55-65. https://doi.org/10.1016/s0894-1777(99)00026-6.

Bucak, H., & Yilmaz, F. (2022). Heat transfer augmentation using periodically spherical dimple-protrusion patterned walls of twisted tape. *International Journal of Thermal Sciences*, *171*, 107211. https://doi.org/10.1016/j.ijthermalsci.2021.107211.

Chavan, H., Tapdiya pa., Birvatkar S., Mundhe S., Sharma V. (2017). Heat transfer enhancement by using a twisted tape insert. *International Journal of Advanced Engineering Research and Studies*, 5, 1152–6.

Chougule, S. S., & Sahu, S. K. (2015). Heat transfer and friction characteristics of Al2O3/water and CNT/water nanofluids in transition flow using helical screw tape inserts–a comparative study. *Chemical Engineering and Processing: Process Intensification*, *88*, 78-88. https://doi.org/10.1016/j.cep.2014.12.005.

Chu, W. X., Tsai, C. A., Lee, B. H., Cheng, K. Y., & Wang, C. C. (2020). Experimental investigation on heat transfer enhancement with twisted tape having various V-cut configurations. *Applied Thermal Engineering*, *172*, 115148. https://doi.org/10.1016/j.applthermaleng.2020.115148.

Chun, B. H., Kang, H. U., & Kim, S. H. (2008). Effect of alumina nanoparticles in the fluid on heat transfer in double-pipe heat exchanger system. *Korean Journal of Chemical Engineering*, *25*, 966-971. https://link.springer.com/article/10.1007/s11814-008-0156-5#citeas.

Córcoles, J. I., Moya-Rico, J. D., Molina, A. E., & Almendros-Ibáñez, J. A. (2020). Numerical and experimental study of the heat transfer process in a double pipe heat exchanger with inner corrugated tubes. *International Journal of Thermal Sciences*, *158*, 106526. https://doi.org/10.1016/j.ijthermalsci.2020.106526.

Dagdevir, T., & Ozceyhan, V. (2021). An experimental study on heat transfer enhancement and flow characteristics of a tube with plain, perforated and dimpled twisted tape inserts. *International Journal of Thermal Sciences*, *159*, 106564. https://doi.org/10.1016/j.ijthermalsci.2020.106564.

Dandoutiya, B. K., & Kumar, A. (2021). CFD analysis for the performance improvement of a double pipe heat exchanger with twisted tape having triangular cut. *Energy Sources, Part A: Recovery, Utilization, and Environmental Effects*, 1-19. https://doi.org/10.1080/15567036.2021.1946215.

Das, A. K., & Hiremath, S. S. (2022). Investigation on the thermohydraulic performance and entropy generation of novel butterfly-wing vortex generator in a rectangular microchannel. *Thermal Science and Engineering Progress*, *36*, 101531. https://doi.org/10.1016/j.tsep.2022.101531.

Dehghani, H., Farhadi, M., Darzi, A. A. R., & Kardgar, A. (2021). A numerical study of flow behavior in the shell and helical finned-tube heat exchanger. *Heat Transfer*, *50*(5), 4607-4621. doi: 10.1002/htj.22091.

Dewan, A., Mahanta, P., Raju, K. S., & Kumar, P. S. (2004). Review of passive heat transfer augmentation techniques. *Proceedings of the Institution of Mechanical Engineers, Part A: Journal of Power and Energy*, *218*(7), 509-527. https://doi.org/10.1243/0957650042456953.

Diwaker, M. K., & Kumar, A. (2023). Impact of cut diameter on thermohydraulic performance of DPHE: an experimental analysis using internet of things (IoT) approach. *Heat and Mass Transfer*, *59*(12), 2299-2310. https://link.springer.com/article/10.1007/s00231-023-03418-z#citeas.

Diwaker, M. K., & Kumar, A. (2023). Thermohydraulic performance of DPHE affected by triangular and semicircular cut size on insert: IoT-based experimentation. *Case Studies in Thermal Engineering*, *43*, 102796. https://doi.org/10.1016/j.csite.2023.102796.

Eiamsa-Ard, S., & Seemawute, P. (2012). Decaying swirl flow in round tubes with short-length twisted tapes. *International Communications in Heat and Mass Transfer*, *39*(5), 649-656. https://doi.org/10.1016/j.icheatmasstransfer.2012.03.021.

Eiamsa-Ard, S., Nivesrangsan, P., Chokphoemphun, S., & Promvonge, P. (2010). Influence of combined non-uniform wire coil and twisted tape inserts on thermal performance characteristics. *International Communications in Heat and Mass Transfer*, *37*(7), 850-856. https://doi.org/10.1016/j.icheatmasstransfer.2010.05.012.

Eiamsa-Ard, S., Wongcharee, K., & Sripattanapipat, S. (2009). 3-D Numerical simulation of swirling flow and convective heat transfer in a circular tube induced by means of loose-fit twisted tapes. *International Communications in Heat and Mass Transfer*, *36*(9), 947-955. https://doi.org/10.1016/j.icheatmasstransfer.2009.06.014.

Elmasry, Y., Chaturvedi, R., Ali, A., Mamun, K., Hadrawi, S. K., & Smaisim, G. F. (2023). Numerical analysis and RSM modeling of the effect of using a V-cut twisted tape turbulator in the absorber tube of a photovoltaic/thermal system on the energy and exergy performances of the system. *Engineering Analysis with Boundary Elements*, *155*, 340-350. https://doi.org/10.1016/j.enganabound.2023.06.013.

Farajollahi, A., Mokhtari, A., Rostami, M., Imani, K., & Salimi, M. (2023). Numerical study of using perforated conical turbulators and added nanoparticles to enhance heat transfer performance in heat exchangers. *Scientia Iranica*, *30*(3), 1027-1038. https://doi.org/10.24200/sci.2022.59717.6394.

Farnam, M., Khoshvaght-Aliabadi, M., & Asadollahzadeh, M. J. (2018). Heat transfer intensification of agitated U-tube heat exchanger using twisted-tube and twisted-tape as passive techniques. *Chemical Engineering and Processing-Process Intensification, 133*, 137-147. https://doi.org/10.1016/j.cep.2018.10.002.

Ganorkar, P. P., & Warkhedkar, R. M. (2015). Heat transfer enhancement in a tube using elliptical-cut twisted tape inserts. *SSRG Int J Mech Eng (SSRG-IJME)*, *2*(5), 20-27.

Gawande, K. R., & Deshmukh, A. V. (2017). Experimental Investigation of Heat Transfer Rate Using Twisted Tape with Elliptical Holes. *IRA-International J Technol Eng(ISSN 2455-4480)*, *7*, 105. https://doi.org/10.21013/jte.icsesd201711.

Goudarzi, K., & Jamali, H. (2017). Heat transfer enhancement of Al2O3-EG nanofluid in a car radiator with wire coil inserts. *Applied Thermal Engineering*, *118*, 510-517. https://doi.org/10.1016/j.applthermaleng.2017.03.016.

Gugulothu, S.K. (2022). Computational fluid dynamics analysis in a ribbed tube with different twisted tape inserts to enhance the heat transfer. *International Journal of Ambient Energy*, *43*(1), 2061-2070. https://doi.org/10.1080/01430750.1722224.

Haddad, F., & Li, P. (2024). Optimization of Dimensions of Smooth and Twisted-Tape-Inserted Tubes for Heat Transfer with NaCl/KCl/MgCl2 Molten Salts by Principle of Entropy Generation Minimization. *Journal of Solar Energy Engineering*, *146*(1). https://doi.org/10.1115/1.4062719.

Hajare, O., Pawar, R., Kadam, J., Kadam, G., & Mokashi, M. (2016). Enhancement of heat transfer with twisted tape inserts and rectangular hole. *International Journal of Current Engineering and Technology, special issue 6.* Available at http://inpressco.com/category/ijcet.

Hamza, N. F. A., & Aljabair, S. (2023, July). Experimental study of heat transfer enhancement using hybrid nanofluid and twisted tape inserts in heat exchangers. In *AIP Conference Proceedings*, 2830(1). AIP Publishing. https://doi.org/10.1063/5.0163912.

Hasanpour, A., Farhadi, M., & Sedighi, K. (2016). Experimental heat transfer and pressure drop study on typical, perforated, V-cut and U-cut twisted tapes in a helically corrugated heat exchanger. *International Communications in Heat and Mass Transfer*, *71*, 126-136. https://doi.org/10.1016/j.icheatmasstransfer.2015.12.032.

Hong, Y., Du, J., Li, Q., Xu, T., & Li, W. (2019). Thermal-hydraulic performances in multiple twisted tapes inserted sinusoidal rib tube heat exchangers for exhaust gas heat recovery applications. *Energy conversion and management*, *185*, 271-290. https://doi.org/10.1016/j.enconman.2019.01.074.

Hong, Y., Zhao, L., Huang, Y., Li, Q., Jiang, J., & Du, J. (2023). Turbulent thermal-hydraulic characteristics in a spiral corrugated waste heat recovery heat exchanger with perforated multiple twisted tapes. *International Journal of Thermal Sciences*, *184*, 108025. https://doi.org/10.1016/j.ijthermalsci.2022.108025.

Hormozi, F., ZareNezhad, B., & Allahyar, H. R. (2016). An experimental investigation on the effects of surfactants on the thermal performance of hybrid nanofluids in helical coil heat exchangers. *International communications in heat and mass transfer*, *78*, 271-276. https://doi.org/10.1016/j.icheatmasstransfer.2016.09.022.

Hwang, S. D., Kwon, H. G., & Cho, H. H. (2010). Local heat transfer and thermal performance on periodically dimple-protrusion patterned walls for compact heat exchangers. *Energy*, *35*(12), 5357-5364. https://doi.org/10.1016/j.energy.2010.07.022.

Ibrahim, S. H., & Abdul Wahhab, H. A. (2022, October). Influence of twisted tape inserts with perforation on heat transfer and pressure drop inside circular tube: Numerical and experimental investigation. In *ICPER 2020: Proceedings of the 7th International Conference on Production, Energy and Reliability* (pp. 281-293). Singapore: Springer Nature Singapore. https://link.springer.com/chapter/10.1007/978-981-19-1939-8_24#citeas.

Jassim, N. A., Hussin, K. A., & Abbass, N. Y. A. (2017). Numerical investigation of heat transfer enhancement in circular tube using twisted tape inserts and nanotechnology. *Wasit Journal of Engineering Sciences*, *5*(2), 42-54. Https://doi.org/10.31185/ejuow.vol5.iss2.57.

Kadbhane, S. V., & Pangavhane, D. R. (2023). Analysis on heat transfer and friction factor characteristics of turbulent flow through heat exchanger fitted with peripheral square-cut twisted tapes arrangement. In *AIP Conference Proceedings* (Vol. 2716, No. 1). AIP Publishing. https://doi.org/10.1016/j.icheatmasstransfer.2013.05.012.

Kadbhane, S. V., & Pangavhane, D. R. (2024). Numerical modeling and optimization of perforated twisted tape insert based on hybrid ANSYS-RSM approach. *Heat Transfer*, *53*(2), 755-785. https://doi.org/10.1002/htj.22971.

Kadhim, Z. K., & Mohammad, S. A. (2017). Experimental Investigation to Enhance the Heat Transfer in Heat Exchanger by Made Groove in Outer Surface of the Inner Tube. *Wasit Journal of Engineering Sciences*, *5*(1), 1-15. https://doi.org/10.31185/ejuow.vol5.iss1.62.

Kadhim, Z. K., & Mohammad, S. A. (2017). CFD Study to Enhance the Heat Transfer in Heat Exchanger by Change the Outer Surface of the Inner Tube and Use Nano Fluid. *Engineering Science*, *2*(3), 58-68. https://doi.org/10.11648/j.es.20170203.12.

Kadhim, Z. K., & Khalaf, K. A. (2020). Numerical study of heat transfer enhancement in contour corrugated channel using water and engine oil. *CFD Letters*, *12*(12), 17-37. ttps://doi.org/10.37934/cfdl.12.12.1737.

Kakaç, S., Liu, H., & Pramuanjaroenkij, A. (2002). *Heat exchangers: selection, rating, and thermal design*. CRC press.

Kamel, H. A., Kadhim, Z. K., & Khalaf, K. A. (2018, March). CFD study of heat transfer enhancement in double pipe heat exchanger with twisted tape inserts. In *2018 International Conference on Advance of Sustainable Engineering and its Application (ICASEA)* (pp. 227-232). IEEE. https://doi.org/10.1109/icasea.2018.8370986.

Kaska, A. S., H Habeeb, H., & A Khalefa, R. (2018). Performance enhancement of the vertical double pipe heat exchanger by applying of bubbling generation on the shell side. *Kirkuk Journal of Science*, *13*(1), 156-171.

Khafaji, H. Q., Abdul Wahhab, H. A., Alsaedi, S. S., Al-Maliki, W. A. K., Alobaid, F., & Epple, B. (2022). Thermal performance evaluation of a tubular heat exchanger fitted with combined basket–twisted tape inserts. *Applied Sciences*, *12*(10), 4807. https://doi.org/10.3390/app12104807.

Khata, N. D., Kadim, Z. K., & Khalaf, K. A. (2021). Numerical and experimental study of heat transfer enhancement in contour corrugated channel using nanofluid and engine oil. In *IOP Conference Series: Materials Science and Engineering* (Vol. 1076, No. 1, p. 012086). IOP Publishing. https://doi 10.1088/1757-899x/1076/1/012086.

Khfagi, A. M., Hunt, G., Paul, M. C., & Karimi, N. (2022). Computational analysis of heat transfer augmentation and thermodynamic irreversibility of hybrid nanofluids in a tube fitted with classical and elliptical-cut twisted tape inserts. *Journal of Thermal Analysis and Calorimetry*, *147*(21), 12093-12110. https://doi.org/10.1007/s10973-022-11418-0.

Kiran, K., Asalammaraja, M., & Umesh, C. (2014). A review on effect of various types of tube inserts on performance parameters of heat exchanger. *International Journal of Research in Advent Technology*, *2*(6), 2321-9637.

Krutova, I., & Zolotonosov, Y. (2020, July). Solution of conjugate problem in a conical coil heat exchanger. In *IOP Conference Series: Materials Science and Engineering* (Vol. 890, No. 1, p. 012156). IOP Publishing. 10.1088/1757-899x/890/1/012156.

Kumar, A., Maithani, R., & Suri, A. R. S. (2017). Numerical and experimental investigation of enhancement of heat transfer in dimpled rib heat exchanger tube. *Heat and Mass Transfer*, *53*(2), 3501-3516. https://link.springer.com/article/10.1007/s00231-017-2080-x#citeas.

Kumar, R., Chandra, P., & -, P. (2020). Innovative method for heat transfer enhancement through shell and coil side fluid flow in shell and helical coil heat exchanger. *Archives of Thermodynamics*, *41*(2), 239-256. 10.24425/ather.2020.133631.

Kumar, R., Kumar, A., Chauhan, R., & Maithani, R. (2018). Comparative study of effect of various blockage arrangements on thermal hydraulic performance in a roughened air passage. *Renewable and Sustainable Energy Reviews*, *81*, 447-463. https://doi.org/10.1016/j.rser.2017.08.023.

Kumar, S., Kothiyal, A. D., Bisht, M. S., & Kumar, A. (2017). Turbulent heat transfer and nanofluid flow in a protruded ribbed square passage. *Results in physics*, *7*, 3603-3618. Https://doi.org/10.1016/j.rinp.2017.09.023.

Legay, M., Simony, B., Boldo, P., Gondrexon, N., Le Person, S., & Bontemps, A. (2012). Improvement of heat transfer by means of ultrasound: Application to a double-tube heat exchanger. *Ultrasonics sonochemistry*, *19*(6), 1194-1200. https://doi.org/10. 1016/j.ultsonch.2012.04.001.

Li, X., Zhu, D., Yin, Y., Tu, A., & Liu, S. (2019). Parametric study on heat transfer and pressure drop of twisted oval tube bundle with in line layout. *International journal of heat and mass transfer*, *135*, 860-872. https://doi.org/https://doi.org/10.1016/ j.ijheatmasstransfer.

Mao, S., Zhou, T., Wei, D., Liu, W., & Zhang, Y. (2021). Modeling of heat transfer of supercritical water in helical finned double pipe. *International Journal of Heat and Mass Transfer*, *180*, 121754. https://doi.org/10.1016/j.ijheatmasstransfer.2021. 121754.

Marzouk, S. A., Abou Al-Sood, M. M., El-Said, E. M., Younes, M. M., & El-Fakharany, M. K. (2023). A comprehensive review of methods of heat transfer enhancement in shell and tube heat exchangers. *Journal of Thermal Analysis and Calorimetry*, *148*(15), 7539-7578. https://doi.org/10.1007/s10973-023-12265-3.

Meyer, J. P., & Abolarin, S. M. (2017). Heat transfer and pressure drop in the transitional flow regime for a smooth circular tube with twisted tape inserts and a square-edged inlet. *International Journal of Heat and Mass Transfer*, *117*, 11-29. https://doi. org/10.1016/j.ijheatmasstransfer.2017.09.103.

Moya-Rico, J. D., Molina, A. E., Belmonte, J. F., Tendero, J. C., & Almendros-Ibáñez, J. A. (2020). Experimental characterization of a double tube heat exchanger with inserted twisted tape elements. *Applied Thermal Engineering*, *174*, 115234. https://doi.org/10.1016/j.applthermaleng.2020.115234.

Mozafarie, S. S., & Javaherdeh, K. (2019). Numerical design and heat transfer analysis of a non-Newtonian fluid flow for annulus with helical fins. *Engineering Science and Technology, an International Journal*, *22*(4), 1107-1115. https://doi.org/10.1016/j.jestch.2019.03.001.

Mogaji, T. S., Kanizawa, F. T., Bandarra Filho, E. P., & Ribatski, G. (2013). Experimental study of the effect of twisted-tape inserts on flow boiling heat transfer enhancement and pressure drop penalty. *International journal of refrigeration*, *36*(2), 504-515.

Mule, B. A., Hatkar, D. N., & Bembde, M. S. (2017). Analysis of Double Pipe heat exchanger with Helical fins. *International Research Journal of Engineering and Technology*, *4*(8), 2395-0072. https://www.academia.edu/download/54374326/irjet-v4i8167.pdf.

Mushatet, K. S. (2023). A review study for a twisted tube heat exchanger. *Journal of Nanofluids*, *12*(2), 299-317. https://doi.org/10.1166/jon.2023.1926.

Mushatet, K. S., & Youssif, F. K. (2018, March). Modeling whirling motion and heat transfer intensification in a heated tube with a twin twisted tape. In *2018 International Conference on Advance of Sustainable Engineering and its Application (ICASEA)* (pp. 215-220). IEEE. https://doi.org/10.1109/icasea.2018.8370984.

Najm, A., Jumaah, I. D., & Karim, A. M. E. A. (2022). Effect of Using A Double Coil Tube with Modified Pitch on The Overall Heat Transfer Rate. *Diyala Journal of Engineering Sciences* Vol., (15) No 1, 40-53. https://djes.info/index.php/djes/article/view/934.

Nakhchi, M. E., & Esfahani, J. A. (2019). Numerical investigation of turbulent Cu-water nanofluid in heat exchanger tube equipped with perforated conical rings. *Advanced Powder Technology*, *30*(7), 1338-1347. https://doi.org/10.1016/j.apt.2019.04.009.

Nakhchi, M. E., & Esfahani, J. A. (2019). Numerical investigation of rectangular-cut twisted tape insert on performance improvement of heat exchangers. *International Journal of Thermal Sciences*, *138*, 75-83. https://doi.org/10.1016/j.ijthermalsci.2018.12.039.

Nakhchi, M. E., Hatami, M., & Rahmati, M. (2020). Experimental investigation of heat transfer enhancement of a heat exchanger tube equipped with double-cut twisted tapes. *Applied Thermal Engineering*, *180*, 115863. https://doi.org/10.1016/j.applthermaleng.2020.115863.

Nakate, M. D. S., Channapattana, M. S., & HC, M. R (2015). Experimental Investigation in the Enhancement of Heat Transfer Rate Using Twisted Tape Turbulators in Heat Exchangers: A Review. *International Journal of Innovative Science, Engineering & Technology*, Vol. 2 Issue 3.

Nalavade, S. P., Deshmukh, P. W., & Sane, N. K. (2022). Heat transfer and friction factor characteristics of turbulent flow using thermally nonconductive twisted tape inserts. *Materials Today: Proceedings*, *52*, 373-378. https://doi.org/10.1016/j.matpr.2021.09.061.

Naphon, P. (2006). Heat transfer and pressure drop in the horizontal double pipes with and without twisted tape insert. *International Communications in Heat and Mass Transfer, 33*(2), 166-175. Https://doi.org/10.1016/j.icheatmasstransfer.2005.09.007.

Naphon, P., Wiriyasart, S., Prurapark, R., & Srichat, A. (2021). Numerical study on the nanofluid flows and temperature behaviors in the spirally coiled tubes with helical ribs. *Case Studies in Thermal Engineering, 27*, 101204. https://doi.org/10.1016/j.csite.2021.101204.

Perng, S. W., Wu, H. W., Kelana, N. P., Guo, Y. L., & Yang, C. J. (2022). Enhancement of turbulent thermal convection in a circular tube with a slotted twisted tape. *International Journal of Numerical Methods for Heat & Fluid Flow, 33*(2), 825-848. https://doi.org/10.1108/hff-05-2022-0288.

Piriyarungrod, N., Kumar, M., Thianpong, C., Pimsarn, M., Chuwattanakul, V., & Eiamsa-ard, S. (2018). Intensification of thermo-hydraulic performance in heat exchanger tube inserted with multiple twisted tapes. *Applied Thermal Engineering, 136*, 516-530. https://doi.org/https://doi.org/10.1016/j.applthermaleng.2018.02.097.

Ponnada, S., Subrahmanyam, T., & Naidu, S.V. (2019). A comparative study on the thermal performance of water in a circular tube with twisted tapes, perforated twisted tapes and perforated twisted tapes with alternate axis. *International Journal of Thermal Sciences, 136*, 530-538. Https://doi.org/10.1016/j.ijthermalsci.2018.11.008.

Prasad, P. D., Gupta, A. V. S. S. K. S., Sreeramulu, M., Sundar, L. S., Singh, M. K., & Sousa, A. C. (2015). Experimental study of heat transfer and friction factor of Al2O3 nanofluid in U-tube heat exchanger with helical tape inserts. *Experimental thermal and fluid science, 62*, 141-150. https://doi.org/10.1016/j.expthermflusci.2014.12.006.

Promthaisong, P., Jedsadaratanachai, W., Chuwattanakul, V., & Eiamsa-ard, S. (2017, August). Simulation of turbulent heat transfer characteristics in a corrugated tube with five-channel twisted tape inserts. In *AIP Conference Proceedings* (Vol. 1879, No. 1). AIP Publishing. Https://doi.org/10.1063/1.5000460.

Rahman, M. T., Habib, K., Quader, M. N., Aslfattahi, N., Kadirgama, K., & Das, L. (2023). Effect of porous density of twisted tape inserts on heat transfer performance inside a closed conduit. *Heliyon, 9*(11). https://doi.org/10.1016/j.heliyon.2023.e21206.

Rahul, N., Kumar, V., Sehrawat, Gupta, T., Manju, R., And Ahmad, M.I. (2017). Design analysis of gasketed plate heat exchanger," *Int. J. Eng. Sci. Res. Technol.*, vol. 4, no. 9, pp. 1682–1685. Https://doi.org/10.3390/en15041340.

Rashidi, S., Esfahani, J. A., & Rashidi, A. (2017). A review on the applications of porous materials in solar energy systems. *Renewable and Sustainable Energy Reviews, 73*, 1198-1210. https://doi.org/10.1016/j.rser.2017.02.028.

Rashidi, S., Eskandarian, M., Mahian, O., & Poncet, S. J. J. O. T. A. (2019). Combination of nanofluid and inserts for heat transfer enhancement: gaps and challenges. *Journal of Thermal Analysis and Calorimetry, 135*, 437-460. https://link.springer.com/article/10.1007/s10973-018-7070-9#citeas.

Rezaei, R. A., Jafarmadar, S., & Khorasani, S. (2019). Presentation of frictional behavior of micro helical tubes with various geometries and related empirical correlation; an experimental study. *International Journal of Thermal Sciences, 140*, 377-387. https://doi.org/10.1016/j.ijthermalsci.2019.03.011.

Saleh, S. A. A., Kadhim, Z. K., & Khalaf, K. A. (2023). An Experimental Study of Heating Heavy Fuel Oil by Hot Air using Helical Fins in a Double-Pipe Heat Exchanger. *Journal of Advanced Research in Fluid Mechanics and Thermal Sciences*, *112*(2), 102-115. https://doi.org/10.37934/arfmts.112.2.102115.

Saleh, S. A., Kadhim, Z. K., & Khalaf, K. A. (2023). CFD simulation of helical coil heat exchanger with Different coil pitch to heating heavy fuel oil. *Wasit Journal of Engineering Sciences*, *11*(3), 120-139. doi: 10.31185/ejuow.vol11.iss3.472.

Salim, K. (2024). A computational study for evaluating the performance of twisted double tube heat exchangers fitted with twisted tape. *CFD Letters*, *16*(5), 21-36. https://doi.org/10.37934/cfdl.16.5.2136.

Salman, S. D., Kadhum, A. A. H., Takriff, M. S., & Mohamad, A. B. (2013). CFD analysis of heat transfer and friction factor charaterstics in a circular tube fitted with horizontal baffles twisted tape inserts. In *IOP Conference Series: Materials Science and Engineering* (Vol. 50, No. 1, p. 012034). IOP Publishing. https://doi.org/10.1088/1757-899x/50/1/012034.

Salman, Z. A., Kadhim, Z. K., & Kamil, H. A. (2021). A Numerical Simulation Study to Improve Heat Transfer Rate in a Double Pipe Heat Exchanger using Different ways. In *Journal of Physics: Conference Series* (Vol. 1973, No. 1, p. 012113). IOP Publishing. https://doi.org/10.1088/1742-6596/1973/1/012113.

Sarafraz, M. M., Hormozi, F., & Nikkhah, V. (2016). Thermal performance of a counter-current double pipe heat exchanger working with COOH-CNT/water nanofluids. *Experimental Thermal and Fluid Science*, *78*, 41-49. https://doi.org/10.1016/j.expthermflusci.2016.05.014.

Shah, R. K. (2003). *Fundamentals of heat exchanger design*. Edited by Ramesh K. Shah, Dusˇan P. Sekulic, John Wiley, and Sons, 2-3. Printed in the United States of America.

Singh, B. P. (2019). A review paper of twisted tape inserts. *International Journal of Applied Engineering Research*, *14*(9), 58-65.

Singh, S. K., Kacker, R., Chaurasiya, P. K., & Gautam, S. S. (2022). Correlations on heat transfer rate and friction factor of a rectangular toothed v-cut twisted tape exhibiting the combined effects of primary and secondary vortex flows. *International Communications in Heat and Mass Transfer*, *139*, 106503. https://doi.org/10.1016/j.icheatmasstransfer.2022.106503.

Singh, S. K., Kacker, R., Gautam, S. S., & Tamang, S. K. (2023). Multi-objective optimization of thermo-hydraulic behavior of heat exchanger with v-cut twisted tape in axial and radial direction using NSGA-II. *Energy Sources, Part A: Recovery, Utilization, and Environmental Effects*, *45*(4), 13057-13076. https://doi.org/10.1080/15567036.2023.2278723.

Singh, S., Pandey, L., Kharkwal, H., & Sah, H. (2020). Augmentation of thermal performance of heat exchanger using elliptical and circular insert with vertical twisted tape. *Experimental Heat Transfer*, *33*(6), 510-525. https://doi.org/10.1080/08916152.2019.1662856.

Sivakumar, K., Rajan, K., Mohankumar, T., & Naveenchnadran, P. (2020). Analysis of heat transfer characteristics with triangular cut twisted tape (TCTT) and circular cut twisted tape (CCTT) inserts. *Materials Today: Proceedings*, *22*, 375-382. https://doi.org/10.1016/j.matpr.2019.07.212.

Sivalakshmi, S., Raja, M., & Gowtham, G. (2021). Effect of helical fins on the performance of a double pipe heat exchanger. *Materials Today: Proceedings*, *43*, 1128-1131. https://doi.org/10.1016/j.matpr.2020.08.563.

Sivashanmugam, P., & Nagarajan, P. K. (2007). Studies on heat transfer and friction factor characteristics of laminar flow through a circular tube fitted with right and left helical screw-tape inserts. *Experimental Thermal and Fluid Science*, *32*(1), 192-197. Https://doi.org/10.1016/j.expthermflusci.2007.03.005.

Sonawane, T., Patil, P., Chavhan, A., & Dusane, B. M. (2016). A review on heat transfer enhancement by passive methods. *International Research Journal of Engineering and Technology*, *3*(9), 1567-1574.

Souayeh, B., Bhattacharyya, S., Hdhiri, N., & Hammami, F. (2021). Numerical investigation on heat transfer augmentation in a triangular solar air heater tube fitted with angular-cut varied-length twisted tape. *The European Physical Journal Plus*, *136*(6), 643. https://link.springer.com/article/10.1140/epjp/s13360-021-01614-6.

Sun, C., Geng, Y., Glowacz, A., Sulowicz, M., Ma, Z., Siarry, P., ... & Li, Z. (2023). The heat transfer and fluid flow characteristics of water/Fe3O4 ferrofluid flow in a tube with V-cut twisted tape turbulator under the magnetic field effect. *Journal of Magnetism and Magnetic Materials*, *586*, 171128. https://doi.org/10.1016/j.jmmm.2023.171128.

Sundar, L. S., & Sharma, K. V. (2010). Turbulent heat transfer and friction factor of Al2O3 nanofluid in circular tube with twisted tape inserts. *International Journal of Heat and Mass Transfer*, *53*(7-8), 1409-1416. https://doi.org/10.1016/j.ijheatmasstransfer.2009.12.016.

Teng, T. P., Hung, Y. H., Teng, T. C., Mo, H. E., & Hsu, H. G. (2010). The effect of alumina/water nanofluid particle size on thermal conductivity. *Applied Thermal Engineering*, *30*(14-15), 2213-2218. https://doi.org/10.1016/j.applthermaleng.2010.05.036.

Thote, S. A., & Singh, N. P. (2023). Analysis of Friction and Heat Transfer Characteristics of Tubes with Trapezoidal Cut Twisted Tape Inserts. *Fluid Dynamics & Materials Processing*, *19*(3). http://dx.doi.org/10.32604/fdmp.2022.021651.

Thote, S. A., & Singh, N. P. (2023). Pipes with Trapezoidal Cut Twisted Tape Inserts in the Laminar Flow Regime: Nusselt Number and Friction Coefficient Analysis. *Fluid Dynamics & Materials Processing*, *19*(2). http://dx.doi.org/10.32604/fdmp.2022.021502.

Thote, S. A., & Singh, N. P. (2024). Improvement of Heat Transfer Through Use of Twisted Tape Inserts: A Review. *Journal of Engineering, Project & Production Management*, *14*(1).

Virgilio, M., Mayo, I., Dedcyne, J., Van Geem, K. M., Marin, G. B., & Arts, T. (2020). Effects of 2-D and 3-D helical inserts on the turbulent flow in pipes. *Experimental Thermal and Fluid Science*, *110*, 109923. https://doi.org/10.1016/j.expthermflusci.2019.109923.

Waghole, D. R., Warkhedkar, R. M., Kulkarni, V. S., & Shrivastva, R. K. (2016). Studies on heat transfer in flow of silver nanofluid through a straight tube with twisted tape inserts. *Heat and Mass Transfer*, *52*, 309-313. https://link.springer.com/article/10.1007/s00231-015-1563-x#citeas.

Wang, Y., Qi, C., Ding, Z., Tu, J., & Zhao, R. (2021). Numerical simulation of flow and heat transfer characteristics of nanofluids in built-in porous twisted tape tube. *Powder Technology*, *392*, 570-586. . https://doi.org/10.1016/j.powtec.2021.07.066.

Wang, Y., Alvarado, J. L., & Terrell Jr, W. (2020). Thermal performance of helical coils with reversed loops and wire coil inserts. *International Journal of Heat and Mass Transfer*, *146*, 118723. https://doi.org/10.1016/j.ijheatmasstransfer.2019.118723.

Wijayanta, A. T., Pranowo, Mirmanto, Kristiawan, B., & Aziz, M. (2019). Internal flow in an enhanced tube having square-cut twisted tape insert. *Energies*, *12*(2), 306. doi:10.3390/en12020306.

Xu, P., Zhou, T., Xing, J., Chen, J., & Fu, Z. (2022). Numerical investigation of heat-transfer enhancement in helically coiled spiral grooved tube heat exchanger. *Progress in Nuclear Energy*, *145*, 104132. https://doi.org/10.1016/j.pnucene.2022.104132.

Yadav, A. S. (2009). Effect of half length twisted-tape turbulators on heat transfer and pressure drop characteristics inside a double pipe u-bend heat exchanger. *JJMIE*, *3*(1), 17-22.

Yakut, K., & Sahin, B. (2004). Flow-induced vibration analysis of conical rings used for heat transfer enhancement in heat exchangers. *Applied Energy*, *78*(3), 273-288. https://doi.org/10.1016/j.apenergy.2003.09.001.

Yildiz, C., Biçer, Y., & Pehlivan, D. (1995). Heat transfers and pressure drops in rotating helical pipes. *Applied energy*, *50*(1), 85-94. https://doi.org/10.1016/0306-2619 (95)90765-9.

Zhang, C., Wang, D., Ren, K., Han, Y., Zhu, Y., Peng, X., ... & Zhang, X. (2016). A comparative review of self-rotating and stationary twisted tape inserts in heat exchanger. *Renewable and Sustainable Energy Reviews*, *53*, 433-449.

Zhang, D., Wu, W., Zhao, L., & Dong, H. (2023). Mathematical Investigation of Heat Transfer Characteristics and Parameter Optimization of Integral Rolled Spiral Finned Tube Bundle Heat Exchangers. *Processes*, *11*(7), 2192. https://www.mdpi.com/2227-9717/11/7/2192.

Zhang, H., Guo, J., Cui, X., & Huai, X. (2020). Performance analysis of supercritical pressure CO2 in several enhanced tubes with non-uniform heat flux. *Applied Thermal Engineering*, *180*,115823. . https://doi.org/10.1016/j.applthermaleng.2020.115823.

Zheng, G., Zhang, W., Man, C., & Sun, P. (2021). Thermal Characteristics of a Heat Exchanger Tube Fitted with Different Peripherally-Cut Twisted Tape Inserts in Laminar Flow. *Heat Transfer Research*, *52*(12). https://doi.org/10.1615/heattransres.2021036646.

Zhu, J. D., & Chen, H. (2015). Numerical study on enhanced heat transfer by twisted tape inserts inside tubes. *Procedia Engineering*, *130*, 256-262. https://doi.org/10.1016/j.proeng.2015.12.219.

Zohuri, B., & Zohuri, B. (2017). Heat exchanger types and classifications. *Compact Heat Exchangers: Selection, Application, Design and Evaluation*, 19-56. https://doi.org/10.1007/978-3-319-29835-1.

Chapter 5

Sustainable Fuel Sourcing for Industrial Oil Burners Through the Use of Biodiesel Blends Made from Waste Cooking Oil

Basil A. Ahmed[1,*]
Adel M. Saleh[1]
and Mohammed A. Fayad[2]
[1]Department of Mechanical Engineering, University of Technology, Baghdad, Iraq
[2]Energy and Renewable Energies Technology Center, University of Technology, Baghdad, Iraq

Abstract

Industrial fossil fuel combustion releases significant amounts of pollutants, causing greenhouse gas emissions, air pollution, and global warming. Several efforts have been made to find sustainable, recyclable, and readily available alternative fuels to reactive burner unit infrastructure adjustments to accomplish sustainable development goals. Sustainability and biodegradability make biodiesel a versatile light diesel oil alternative. Multiple studies have examined the effects of employing a Biodiesel-Diesel blend with Exhaust Gas Recirculation (EGR) on burner efficiency and emitted pollutants. By adding biodiesel to diesel fuel blends at specific ratios, its performance improves, allowing it to meet international pollution standards. As bio-diesel mixes grow, CO emissions fall. Biofuels reduce CO_2 emissions due to their oxygen content. Increased oxygen combustion reduces unburned hydrocarbons.

* Corresponding Author's Email: basil.basil2003@gmail.com

In: Advancements in Sustainability Systems II
Editors: H. H. Al-Kayiem, R. A. Jessam, Hasanain A. A. Wahhab et al.
ISBN: 979-8-89530-562-1

Biodiesel and biodiesel mix with additives change when exposed to high temperatures. Effective EGR reduces NOx emissions, available oxygen emissions, and flame temperature while raising CO2 concentration, according to research. This lowers hazardous Biodiesel emissions in burner systems. This review study examines biodiesel production methods in detail. The most promising diesel fuel replacement is biodiesel. Due to its renewable nature, current manufacturing method, sustainability potential, economic viability, high combustion efficiency, biodegradability, environmental friendliness, and similar chemical and physical qualities to diesel oil. This article summarizes biodiesel efficiency and environmental impact studies. This review suggests that biodiesel fuel could be a sustainable burner fuel source. The purposeful modification of EGR rates shows its effectiveness in balancing biodiesel's good environmental advantages with emissions management and reduction.

Keywords: biodiesel, waste cooking oil, exhaust gas recirculation, emissions and performance of burner, oil burner, swirler burner

Abbreviations

EGR	Exhaust Gas Recirculation
NOx	Nitrogen oxide
O_2	Oxygen
CO	Carbon monoxide
CO_2	Carbon dioxide
SO_2	Sulfur dioxide
UHC	unburned hydrocarbons
JOME	Jatropha oil methyl ester
COME	Coconut oil methyl ester
POME	palm oil methyl ester
WCO	Waste oil cooking
LDO	Light diesel oil
FGR	Flue Gas Recirculation
DD	Distillate diesel
FGB	First Generation Biodiesel
SGB	Second Generation Biodiesel
FIR	Flue-gas internal recirculation

1. Introduction

To achieve sustainable development objectives, it is essential to decrease emissions of carbon dioxide (CO2), carbon monoxide (CO), unburned constituents, particulate matter (PM), nitrogen oxides (NOx), and sulfur oxides (SOx) from oil-based industrial burners. According to projections, the demand for modern energy sources will increase in tandem with industrial advancements. The industrial sector has experienced a 49% increase in CO2 emissions as a consequence of the utilization of fossil fuels. The energy sector reduces CO2 emissions by utilizing renewable energy sources (Adhikesavan et al. 2022; Razzaq et al. 2020; Suzihaque et al. 2022).

1.1. Background of Biodiesel

Biodiesel is seen as a highly promising substitute for conventional diesel fuel. Biodiesel offers advantages over diesel fuel in terms of increased cetane numbers, reduced sulfur and aromatics levels, and higher flash and cloud points. These benefits have a good impact on the combustion properties, but the fuel lacks oxidation stability and results in greater levels of NOx and lower levels of CO and particulate matter (PM) emissions (Fathurrahman et al. 2021; Kass et al. 2022). Biodiesel is a sustainable and durable alternative to conventional fuel that has attracted significant worldwide attention. Biodiesel is produced through the transesterification process of vegetable oils or animal fats, resulting in a fuel consisting of mono-alkyl esters. Biodiesel has several advantages, including its inherent biodegradability, renewable characteristics, non-toxic nature, and reduced pollution levels in comparison to fossil fuels. Biodiesel production can employ a variety of feedstocks, including both edible and non-edible oils. Palm, soybean, rapeseed, sunflower, Canola, corn, and coconut are examples of edible oils, while non-edible oils are derived from sources including jatropha, castor, Karanja, Pongamia, and rubber. Jatropha oil is an economically efficient source of biodiesel that contains a large amount of oil and has the potential for high production of seeds (Gad et al. 2021; Norwazan et al. 2018).

Multiple research publications have examined the combustion and emission characteristics of biodiesel and diesel oil mixes, focusing on reducing emissions and improving combustion efficiency. An investigation was conducted to assess biodiesel's efficiency and environmental impact, and its mixtures with diesel fuel when utilized in boilers. Studies have shown that diesel fuel is somewhat more efficient than biodiesel when used at higher energy levels, but biodiesel is more efficient than diesel at lower energy levels. Regarding emissions, biodiesel and its blends generally emit lesser quantities of CO, SO_2, and CO_2 than diesel. However, it is worth noting that B10, a blend of biodiesel and diesel, does release higher levels of (SO_2).

Furthermore, the combustion of biodiesel has been found to reduce the quantity of NO_x emissions in flue gases (Ghorbani et al. 2011). An assessment was conducted on the combustion efficiency of biodiesel and diesel oil mixes using a liquid fuel burner. The biodiesel fuels were derived from several plant oil sources, including jatropha and palm, and coconut oils. The findings indicated that the combination of jatropha oil methyl ester (JOME) and coconut oil methyl ester (COME) at a volume ratio of 25% in diesel fuel resulted in reduced emissions of carbon monoxide (CO) and unburned hydrocarbons (UHC), suggesting improved combustion efficiency. However, when a blend of palm oil methyl ester (POME) at a proportion of 25% was used, there was an observed increase in emissions of carbon monoxide (CO) and unburned hydrocarbons (UHC). The combustion of all biodiesel blend fuels resulted in slightly elevated nitrogen oxide (NO_x) emissions compared to pure diesel fuel combustion.

Nevertheless, all blends of biodiesel exhibited a substantial reduction in emissions of SO_2 and unburned hydrocarbons (UHC). In summary, biodiesel-diesel oil blends are a suitable choice for liquid fuel burners as they result in reduced emissions and improved environmental impact (Kumaran et al. 2014). Previous studies by Mahfouz et al. 2018 have indicated that combining waste cooking oil (WCO) with light diesel oil (LDO) can lead to satisfactory efficiency of the combustor and reduced emissions. It is recommended to blend up to 20% WCO with LDO.

Furthermore, the use of $A_{l2}O_3$ nanoparticles in B20 blended biodiesel fuel has been discovered to improve flame properties and thermal efficiency and decrease the release of pollutants (Pourhoseini & Ghodrat, 2021). Research has also examined the use of biodiesel made from palm oil. It has been found that increasing levels of biodiesel content result in faster combustion rates and increased emissions, except for NOx (Ahmad et al. 2020). In addition, the burning of cottonseed oil in a modified burner has demonstrated appropriate spray conditions and emissions like heating fuel oil (Daho et al. 2014). Investigating biodiesel blend mixture development in burner systems aims to get insight into its impact on combustion and emissions. Research has shown that adding water to biodiesel blends enhances the mixing of fuel and air, as well as the penetration of the fuel spray, resulting in decreased emissions of CO, CO_2, NO_x, and HC (Khalid et al. 2014). Research has been conducted using biodiesel-diesel oil blends made from waste cooking oil, as summarized in Table 1. This research focuses on regulating the combustion process to reduce the release of harmful emissions (Suardi et al. 2006).

Table 1. The main methods and outcomes for Diesel/Waste cooking oil blend

Name of authors	Investigation Method(s)	Remarks
Gad et al. 2021	• The compressor and blower control the supply of primary and secondary air. • The efficiency of the combustor is determined by calculating it using thermocouples.	• The exhaust gas temperature fluctuated depending on the blends of WCO, but the emissions were reduced compared to diesel fuel. • Combustor efficiency is reduced when using WCO mixes compared to diesel. • The levels of O2 and SO2 emissions fluctuated depending on the mixtures of blended oil and diesel oil. • Uncombusted hydrocarbons and NOx emissions are influenced using WCO blended with diesel oil.
El-Shenawy et al. 2018	• An experimental device used to test the burning process and investigate the resulting emissions. • Analysis of combustion through fuel blending, mixing, and recirculation. • The scope and precision of instruments used for gas analysis. • Continuous flame supply system for combining air and oxygen	• The emission of carbon monoxide reduces as the quantity of biodiesel in blends increases. • Biofuels decrease CO2 emissions because of their elevated oxygen content. • The presence of higher oxygen levels in combustion leads to a reduction in unburned hydrocarbons.
Hosseini et al. 2010	• An empirical study on the combustion process and emission of contaminants from biodiesel and oil gas.	• Biodiesel improves the combustion process and decreases emissions of CO, CO2, and particulate matter. • The use of biodiesel leads to an increase in NOx emissions, but there is a decrease in CO and CO2 emissions.

In general, using biodiesel-diesel oil mixtures could decrease emissions and enhance burner systems' performance. Flue Gas Recirculation (FGR) is a method employed to decrease NO_x emissions and enhance the efficiency of combustion systems. The process entails redirecting a fraction of the exhaust gas back into the combustion chamber, which reduces flame temperature and restricts the amount of oxygen available for nitrogen oxidation. This leads to a decrease in the conversion of nitrogen bound in fuel and the generation of thermal NO_x. FGR can be applied in several combustion systems, including wet-bottom boilers and installations that use oil or gas as fuel. The quantity of re-circulated flue gas is generally restricted to approximately 30% to prevent operational constraints and minimize efficiency reductions. Advanced low-NO_x burners can be employed to offset the elevated levels of NOx emissions. In summary, FGR is a highly efficient technique for decreasing NO_x emissions in combustion systems operating at high temperatures (Tanasić et al. 2022).

1.2. Background and Motivation of Flue Gas Recirculation

FGR is a method that has been suggested and studied for its impact on emissions and efficiency in natural gas burners. FGR, or Flue Gas Recirculation, improves the brightness of the natural gas flame, expands the surface area for heat transfer, and removes areas of high temperature, reducing the flame's temperature along its axis and minimizing thermal corrosion. FGR also amplifies the level and duration of significant irradiative substances in the flame, intensifying the infrared emissivity and heat transfer by radiation (Pourhoseini et al. 2021). Because of the properties of the fuel utilized and the unique air-fuel mixture that is generated, the combustion characteristics of blends and single fuels would differ during burner operation. The air-fuel mixture would directly influence the performance of the burner and the pollutants generated. The methods proposed so far for the flue gas recirculation in burners are summarized in Table 2.

Additionally, re-circulated exhaust gas can be combined with new combustion air in an industrial furnace equipped with exhaust gas recirculation, which reduces NOx emissions and fuel consumption (Ciupek et al. 2024; Kikuchi et al. 2022). In addition, FGR lowers the flue gas temperature at the furnace's exhaust, minimizing energy loss through the chimney (Kwon et al. 2018). FGR is a highly efficient method for decreasing NO_x emissions in combustion systems. This process entails reintroducing flue gas from the boiler's exhaust into the oxidizer, decreasing the oxygen concentration and maximum temperature of the flame in the combustion area. As a result, the area with high temperature becomes smaller, and the temperature in the combustion zone becomes more consistent (Wang et al. 2018). The process reintroduces flue gas from the boiler's tail into the oxidizer. This reduces the concentration of oxygen and the highest temperature of the flame, leading to decreased NO_x emissions and improved temperature consistency in the combustion area (Abdul Rahim et al. 2016).

FGR can be integrated with additional combustion techniques, such as fuel/air staging and lean-burn premixed combustion, to enhance emissions reduction efficiency, as shown in Figure 1. Nevertheless, it is imperative to regulate the FGR rate to fall within 10% to 20% to prevent unstable combustion and flameout, as stated in reference (Tanasić et al. 2016). Research has demonstrated that implementing FGR can substantially decrease (NOx) emissions, with reductions of up to 80% compared to systems that do not utilize FGR. Optimizing the design and operation of FGR systems is crucial to achieving the desired reduction in emissions while ensuring stable combustion (Zhu et al. 2021). To maintain a consistent combustion flame, it is recommended to keep the present FGR rate for industrial boilers between 10% and 20% (Li et al. 2013). By combining FGR with additional combustion methods like fuel/air staging, the decrease of NO_x emissions can be further improved (Saraçoğlu & Gündüz, 2009).

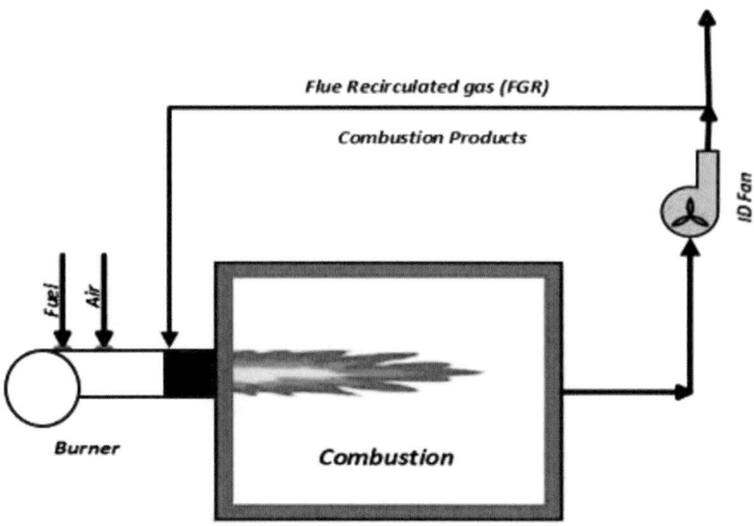

Figure 1. FGR system with burner.

In summary, FGR is a highly promising technique for mitigating emissions in combustion systems. Nevertheless, achieving optimal performance requires diligent monitoring and meticulous fine-tuning. Implementing strategic exhaust gas recirculation rates offers a viable approach to balancing the environmental benefits of using biodiesel with the necessity of regulating and reducing harmful emissions. Hence, the interaction between the rate of the EGR technique and the different percentages of biodiesel blend is crucial to consider when aiming to enhance the performance and the emission characteristics of the burner.

Table 2. The main methods and outcomes for FGR

Name of authors	Investigation Method(s)	Remarks
Tanasić et al. 2022	• The flue gas recirculation system was evaluated to determine its effectiveness in reducing NOx emissions. •Numerical research, chemiluminescence imaging, and industrial experiments were performed. •A gas analyzer is utilized to analyze the components of NOx.	• NOx emissions decreased by 20-50% •Flue gas recirculation efficiently reduced NOx emissions in utility boilers. •Hot water boilers are now operating at levels near the desired values, indicating the possibility for further enhancement. • The FGR system decreased the amount of oxygen accessible, resulting in a decrease in the generation of thermal NOx. • The boilers emitted NOx in the range of 85.3 to 108.4 mg/m3.
Zhu et al. 2021	• The effectiveness of burners was assessed by industry experiments and computational fluid dynamics (CFD) simulations. •The model includes transport equations for the variables k and e. • Reaction coefficients and equations for the rate of thermal NO production are included.	• The numerical results have been verified using experimental data, demonstrating a high level of agreement. •The ideal excess air coefficient for achieving minimal NO emissions is 1.15. • There are no spikes in the concentration of NO (nitric oxide) seen downstream of the flame, either in the air or on the surface.

2. Methodologies of Biodiesel Production

Researchers are encouraged to utilize alternative fuels, like biodiesel, because there is a limited supply of diesel fuel around the globe and because environmental protection agencies are working hard to minimize its utilization (Huang et al. 2022; Pandey & Murugan, 2022; Sun et al. 2020). The process of preparing biodiesel involves several sequential processes, as shown in Figure 2. Initially, biodiesel, a renewable alternative fuel, is produced using raw materials such as vegetable oils or animal fats. Various varieties of oils, including soybean oil, rapeseed oil, coconut oil, and palm oil, have been employed as primary resources for the production of Biodiesel (Manikandan et al. 2023). After the previous phase, the next step is transesterification, which converts the raw material into biodiesel. This technique is crucial when the oil's free fatty acid content exceeds 1% by weight. Through a chemical reaction involving alcohol, typically methanol, and a catalyst, the triglycerides are mixed to produce biodiesel. A catalyst is essential for increasing both the pace of reaction and the amount of biodiesel produced (Li et al. 2013; Ng and Gan, 2010). The production and conversion of biodiesel are affected by multiple factors, including the kind and amount of variables such as feedstock, alcohol, molar ratio, catalyst, reaction temperature, and duration (Singh & Singh, 2010; Tanasić et al. 2022). It is possible to collect the waste cooking oil from the factory and then heat it to a temperature of 95°C. Subsequently, the liquid is subjected to a mesh filter to remove larger particles, followed by filtering using a cartridge filter to eliminate impurities. Once the oil has undergone filtration, it is prepared for combustion (Mahfouz et al. 2018; Atadashi et al. 2012). When biodiesel is used, it produces lower levels of pollutants such as nitrogen oxides and smoke opacity compared to diesel fuel.

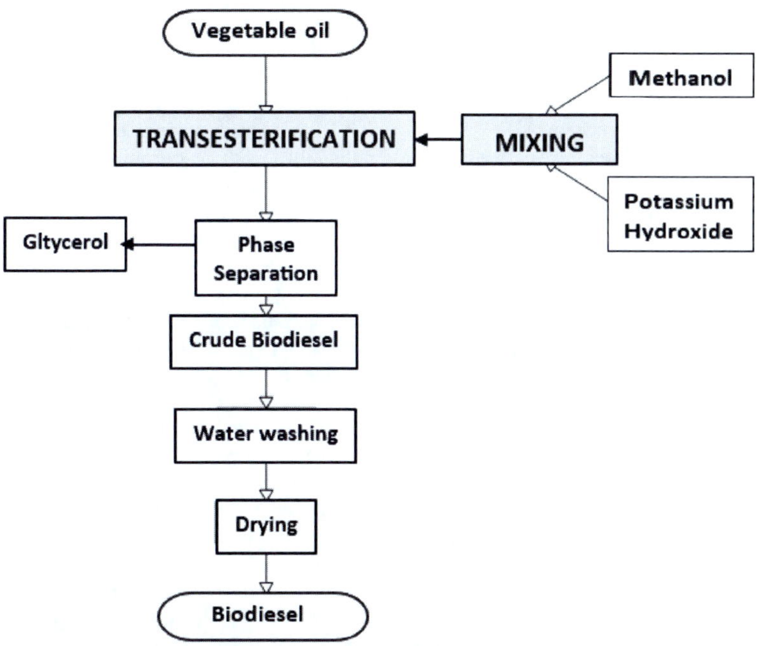

Figure 2. Scheme for preparing a type of biodiesel fuel.

3. Impact of Biodiesel Blend on the Performance and Emissions of Burners

Biodiesel blends can reduce sulfur dioxide (SO_2) and UHC emissions compared to diesel fuel (El-Shenawy et al. 2018; Hashimoto et al. 2014). The reduction in NO_x emissions for biodiesel blends is attributed to improved atomization characteristics and fuel properties (Li et al. 2013). Overall, the use demonstrates the potential of biodiesel/diesel blends to improve burner performance and reduce emissions. Table 3 shows the main results with methods for biodiesel.

Table 3. The main results with methods for biodiesel

Name of authors	Investigation Method(s)	Remarks
Ghorbani et al. 2011	•The process of converting vegetable oils into methyl ester through transesterification, utilizing potassium hydroxide. • Comparison of combustion efficiency and flue gas emissions in an experimental boiler. •Utilization of soybean and sunflower oil in the creation of biodiesel	• Biodiesel blends produced lower emissions of carbon monoxide (CO), sulfur dioxide (SO2), and carbon dioxide (CO2) compared to diesel fuel. • Biodiesel combustion at the second energy level leads to a decrease in NOX emissions. • Diesel efficiency is marginally superior to biodiesel, with the exception of situations with lower energy levels.
Mahfouz et al. 2018	• The compressor and blower control the supply of primary and secondary air. • The efficiency of the combustor is determined by measuring temperatures with thermocouples.	• The exhaust gas temperature changed depending on the blends of waste cooking oil (WCO), and the emissions decreased compared to diesel fuel. • The combustor efficiency is reduced when using WCO mixes compared to diesel. • The levels of oxygen and SO2 emissions changed depending on the mixtures of waste cooking oil and diesel oil. • The use of WCO-diesel oil mixtures influences the emissions of unburned hydrocarbons and NOx.
Khalid et al. 2014	• Optical visualization technology is used to analyze the creation of mixtures and the production of emissions. • An auto-check 5-channel gas emission analyzer is used to measure exhaust emissions.	• Augmenting the water content decreases exhaust emissions, particularly NOx and HC. • Water emulsification improves the process of mixing air and fuel, leading to increased efficiency in combustion. • The utilization of waste cooking oil for biodiesel production has an impact on the creation of fuel mixtures and the resulting emissions

Biodiesel blends have been investigated for their impact on burner performance and emissions. Several studies have focused on this topic. Khalid et al. 2014 conducted experimental work on the combustion of biodiesel blends in a liquid fuel burner. The blends produced lower carbon monoxide (CO) and unburned hydrocarbon (UHC) emissions than pure diesel fuel. They also observed slightly higher nitrogen oxide (NO_x) emissions for the biodiesel blends. Similarly, another study by Khalid et al. examined the effects of biodiesel fuel from waste cooking oil on mixture formation and emissions in a burner system. They found that increasing the percentage of water content in the biodiesel blend reduced exhaust emissions of CO, CO_2, NO_x, and UHC. The emissions curves of (WOC) are present in Figure 3 (Anand et al. 2011; Kumaran et al. 2014; Tan et al. 2011). Biodiesel blend fuels from different plant oil feedstocks, such as jatropha, palm, and coconut oils, were also investigated. It was found that these biodiesel blends produced lower CO and UHC emissions but slightly higher NO_x emissions compared to diesel fuel (Abdul Rahim et al. 2016, Rehman et al. 2011, and Sharma & Singh, 2008). Oxy-fuel combustion with flue gas recirculation was another method studied to reduce emissions. It was found that oxy-fuel combustion with FGR reduced NO_x and SO_2 emissions but slightly increased CO_2 emissions (Tanasić et al. 2016). The emissions of NO_x from some biodiesel blends have decreased in comparison with fossil diesel, except for certain blend percentages (Sharon et al. 2012; Zhu et al. 2021).

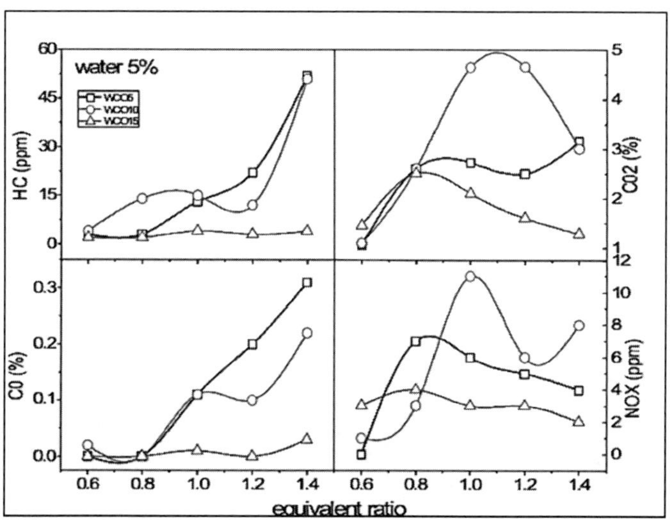

Figure 3. (Continued).

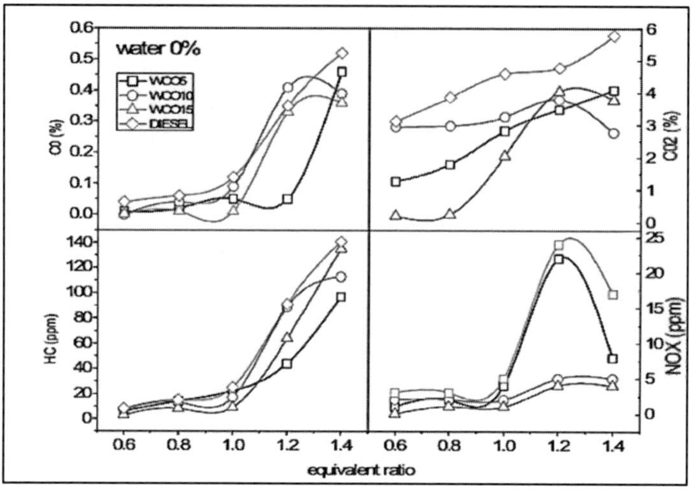

Figure 3. The effect percentage of water and the equivalent ratio for WCO biodiesel fuel on emissions (Daho et al. 2014).

4. Effect of Flue Gas Recirculation on Emissions and Performance Characteristics of Burner

The performance characteristics of using flue gas recirculation on a burner can be observed regarding combustion efficiency, flame stability, and emissions reduction. FGR has been shown to improve the physical properties of biodiesel, such as viscosity and density, leading to better atomization and combustion characteristics (Kumaran et al. 2014). In the case of a low NO_x burner for a gas boiler, FGR has been shown to reduce NO_x emissions and increase flame radiation heat flux (Pourhoseini et al. 2021). Similarly, in natural gas burners, tangential flue gas regeneration TFGR has been found to enhance flame radiation characteristics and reduce NOx emissions (Kwon et al. 2018). Additionally, FGR can enhance convective heat transfer and reduce radiation heat loss, resulting in improved system performance (Tanasić et al. 2016). Flue gas recirculation in oxy-fuel combustion, which uses an oxidizer mixture mainly composed of oxygen and carbon dioxide, reduces unburned fuel and lowers emissions, including NO_x and increased carbon dioxide concentration (Chen et al. 2017). The use of FGR in oxy-fuel combustion has been found to reduce NO emissions and increase CO_2 concentration, making it a suitable method for carbon capture and storage (Zhu et al. 2021). Flue gas recirculation can significantly impact the combustion efficiency of biodiesel blends. It has been observed that increasing the recirculation ratio of flue gas can lead to a reduction in NO_x emissions, which is beneficial for improving combustion efficiency. The recycled flue gas dilutes the oxygen concentration in the combustion zone, resulting in a lower peak flame temperature and a more uniform temperature distribution. This helps to delay and distribute the heat-releasing reactions of the fuel and oxidizer, leading to a more efficient combustion process.

Additionally, flue gas recirculation can also contribute to the reduction of thermal NO_x formation, further enhancing combustion efficiency (Khalid et al. 2014). As the recirculation ratio increases, NO_x emissions decrease due to oxygen dilution in the combustion zone, resulting in lower peak flame temperatures and reduced thermal NO_x formation (Li et al. 2013). The NOx emissions curves are present in Figure 4. Overall, using FGR in burners can improve combustion performance and reduce emissions. Table 4 shows the main results with methods for FGR.

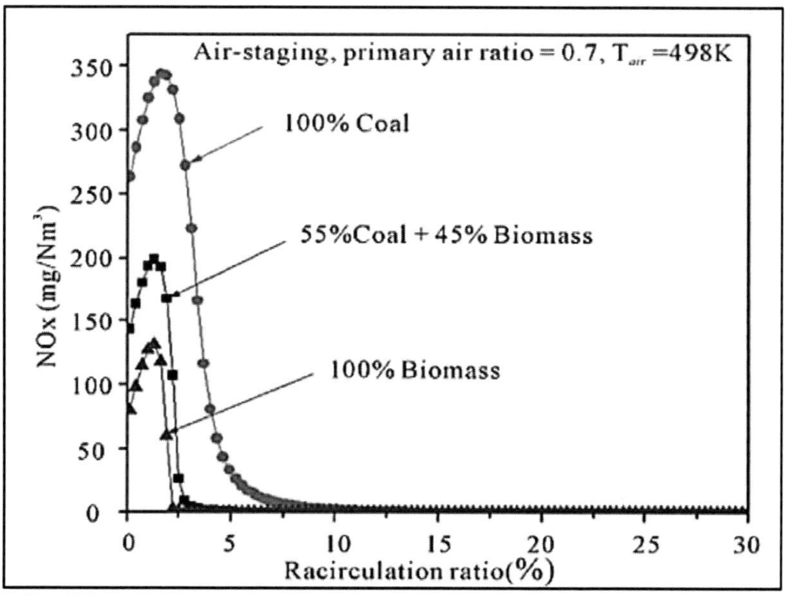

Reproduced from Tanasić et al. 2022.

Figure 4. Effect of different EGR with fuel types on NO_x emission.

Table 4. The main results with methods for FGR

Name of authors	Investigation Method(s)	Remarks
Li et al. 2013	• Aspen Plus model for a combustion system with volume-based measurements. • Elaborate models for the process of devolatilization, combustion, and flue gas recirculation. •Study of flue gas recirculation, air preheating, oxygen content, and fuel kinds. • Investigation of NOx and SOx emissions under various circumstances	•Increasing the biomass ratio in co-firing boilers leads to a decrease in NOx and SOx emissions. Flue gas recirculation reduces nitrogen oxide (NOx) emissions, but it also leads to a minor rise in sulfur oxide (SOx) emissions. • Biomass combustion requires less energy than coal to achieve low levels of NOx emissions.
Nhan et al. 2019	• An experimentally confirmed response rate model has been enhanced with a position-dependent scaling factor. •Flue gas return is quantified using FIR efficiency and FIR fraction. • Thorough computational fluid dynamics (CFD) simulations performed using an innovative low-NOx burner design	• The FIR burner effectively decreases NOx emissions in comparison to a non-FIR system. • The selected model for CFD analysis is used to study the FIR system.
Kwon et al. 2018	• Conducting a numerical study on the reduction of NOx in a combustion system by recirculation. • Examination of the impacts of buoyancy and heat radiation on combustion systems. • Utilization of computational fluid dynamics to precisely forecast NOx emissions. • Introducing a low-NOx burner that utilizes flue-gas internal recirculation	• Flue-gas recirculation resulted in a 50% reduction in NOx emissions due to the decrease in temperature. • The presence of buoyancy increased NOx levels in the non-recirculation system but decreased them in the recirculation system. • Buoyancy and thermal radiation are essential factors for accurately predicting NOx emissions.
Pourhoseini et al. 2021	• Measuring the temperature of a flame using an S-type thermocouple. •Optical characteristics were examined using a digital brightness meter and infrared photography. • The SBG01 heat flux sensor was used to detect the overall radiation.	• TFGR improves flame brightness by causing the dissociation of carbon monoxide, resulting in a decrease in nitrogen oxide (NOx) emissions. • TFGR enhances the concentration of carbon monoxide (CO), the emissivity of infrared (IR) radiation, and the transmission of heat through radiation. •The flame radiation heat flux has a 74% rise when the TFGR ratio is set to 70.

5. Biodiesel Blends and Exhaust Gas Recirculation in Industrial Burners

1. Biodiesel is a sustainable and sophisticated fuel that decreases exhaust emissions and reduces reliance on imported crude oil. The characteristics of biodiesel are crucial for its application in diesel engines or furnaces. The EGR system is installed in several engines and burners to minimize exhaust emissions by utilizing internal exhaust gas (Elkelawy et al. 2023; Gaur & Goyal, 2022; Linganiso et al. 2022; Milano et al. 2022; Samuel et al. 2020, 2021; Simsek, 2020; Yilmaz et al. 2022).
2. EGR gas typically contains an oxygen percentage ranging from 15 to 20%, along with other exhaust gas components like hydrocarbons, carbon monoxide, nitrogen oxides, and a lower quantity of particulate matter. The EGR components are inexpensive, and the technology utilized in the EGR is straightforward. The EGR has a minimal impact on other pollutants, with the exception of particulate matter, when it comes to decreasing system NOx (Yu et al. 2022; Yusaf et al. 2022; Dey et al. 2021; Kulkarni et al. 2023; Mizik & Gyarmati, 2021; Pan, 2021; Prasad et al. 2020; Sentanuhady et al. 2022; Yusoff et al. 2021).
3. The EGR can be employed in conjunction with an oil burner to analyze burner efficiency, emissions, and flame properties. The EGR system can be implemented in an oil burner, as shown in Figure 5. In that case, it has the potential to gradually replace burner technologies such as Selective Catalytic Reduction (SCR), Selective Non-Catalytic Reduction (SNCR), and Overfire Air (OHCR). The EGR can also serve as an alternative control method to these more expensive systems. Presently, the utilization of EGR technology with oil burners is not prevalent in industrial thermal applications. There is a lack of literature regarding the impact of biodiesel mixes with EGR on the performance of industrial oil burners, as well as emissions of nitrogen oxides (NOx), carbon monoxide (CO), unburned hydrocarbons, and particulate matter. Nevertheless, there is existing research that discusses the impact of biodiesel mixes with exhaust EGR on the performance of cottonseed oil burners, emissions, and the quality of the flame (Alsultan et al. 2020; Ershov et al. 2023; Esakki et al. 2022; Kamaraj & Rao, 2022; Kaya & Kökkülünk, 2023; Palani et al. 2022; Patil et al. 2024; Yesilyurt et al. 2020; Yesilyurt & Aydin, 2020).

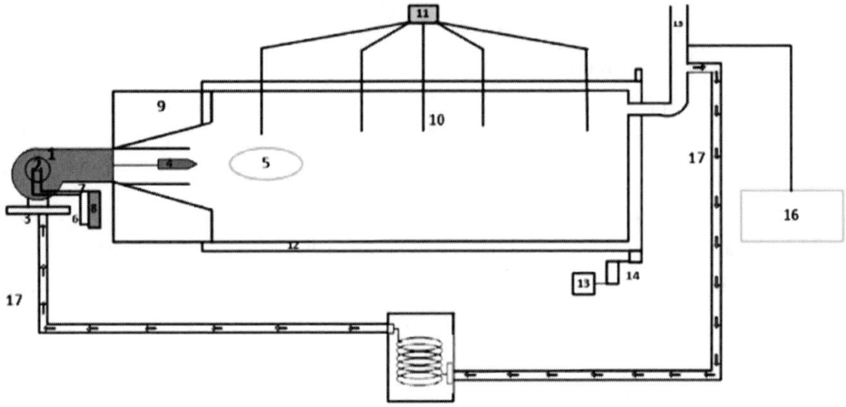

No	Description	No	Description	No	Description
1	Burner	7	Bypass fuel	13	Water pump
2	Fuel pump	8	Fuel tank	14	Water flow meter
3	Air flow meter	9	Insulation block	15	Chimney
4	Nozzle	10	Thermocouple	16	Gas analyzer
5	Flame	11	Thermometer	17	EGR system
6	Fuel suction	12	Cooling water		

Source: Authors.

Figure 5. The schematic diagram of the experimental rig.

6. Economic Analysis of Converting Waste Cooking Oil to Biodiesel

The conversion of waste into biodiesel represents a renewable energy alternative. Used cooking oil constitutes a significant problem. It results in fatbergs, pipeline obstructions, and subterranean water contamination. The increase in fast food consumption resulting from economic growth complicates waste oil disposal for micro and small eateries. Waste can be utilized to augment finite resources, such as food, if addressed effectively. Reusing waste conserves financial resources and enhances advantages. The utilization of waste resources is cost-effective and yields significant economic benefits. These advantages encompass reduced gasoline expenses, enhanced foreign exchange earnings, and diminished CO2 emissions. Its waste origin allows for the conservation of primary energy. Additional advantages encompass ecological sustainability and economic viability. Consequently, utilizing this waste for energy seems advantageous. Energy is fundamental to economic development and well-being (Adewuyi, 2020; Athew et al. 2021).

Consequently, the economic analysis has as its objective to verify the possibility of carrying out the process from the economic point of view (cost analysis), to understand the economic sustainability concerning the market (price analysis), and to verify the effect of government policies on the convenience of investing in this kind of process(Subbaramaiah & Sujatha, 2023). Internal economic variables that may affect the costs of production are, for example, the cost of the feedstock, processing costs, depreciation, financing costs, taxation, and profit to the owner of the plant. The market prices of both feedstocks and the final product are two important variables influencing the economic sustainability of a biodiesel project(Sarwer et al. 2022; Suhara et al. 2024). the use of waste cooking oil as a feedstock is economically viable and can drive long-term sustainability through a reduction in both energy demand and greenhouse gas emissions. Moreover, producing biodiesel from waste cooking oil supports the production and utilization of alternative energy sources and is beneficial to public health (Manikandan et al. 2023).

Conclusion

Studies reveal combustion, emissions, and NOx reduction methods. We stressed essential elements like:

A. Adding biodiesel to blends reduces carbon monoxide emissions due to higher oxygen content and increased oxygen rate in combustion.
B. FGR system reduces NO_x emissions and available oxygen emissions and lowers flame temperature with increased CO2 concentration.
C. Biodiesel as fuel increases NO_x emissions, while CO and CO_2 emissions decrease when compared to using gas oil as fuel.
D. Internal flue gas recirculation reduces NOx emissions but may increase SOx emissions. Air preheating temperature and oxygen concentration have little effect on NOx emissions when the recirculation ratio is large.
E. The CFD simulations of the FIR burner assess its ability to reduce NOx emissions. Hot flue gas is drawn back to heat cold air in the FIR burner, lowering NOx emissions.

F. A study on biodiesel blends and exhaust gas recirculation found that raising the air-to-fuel ratio lowers the flame temperature, and the biodiesel blend has a lower heating value than pure diesel.

These findings emphasize the relevance of burner design, combustion technology, and fuel mix in reducing NOx emissions and improving combustion efficiency.

References

Abdul Rahim, N., Mohd Jaafar, M. N., Sapee, S., & Elraheem, H. F. (2016). Effect on particulate and gas emissions by combusting biodiesel blend fuels made from different plant oil feedstocks in a liquid fuel burner. *Energies*, 9(8), 659. https://doi.org/10.3390/en9080659.

Adewuyi, A. (2020). Challenges and prospects of renewable energy in Nigeria: A case of bioethanol and biodiesel production. *Energy Reports*, 6, 77–88.

Adhikesavan, C., Ganesh, D., & Augustin, V. C. (2022). Effect of quality of waste cooking oil on the properties of biodiesel, engine performance and emissions. *Cleaner Chemical Engineering*, 4, 100070.

Ahmad, A. A., Jaafar, M. N. M., Othman, N. B., Azli, A. A. M., Said, M., Rahim, M. R., Malik, M. S. A., & Omar, W. Z. W. (2020). Combustion and Emission Characteristics of Palm Oil-Based Biodiesel in a Liquid Fuel Burner. *IOP Conference Series: Materials Science and Engineering*, 884(1), 12026. https://doi.org/10.1088/1757-899X/884/1/012026.

Alsultan, A. G., Asikin Mijan, N., Mansir, N., Razali, S. Z., Yunus, R., & Taufiq-Yap, Y. H. (2020). Combustion and emission performance of CO/NOx/SOx for green diesel blends in a swirl burner. *ACS Omega*, 6(1), 408–415. https://doi.org/10.1021/acsomega.0c04800.

Altarazi, Y. S. M., Yu, J., Gires, E., Ghafir, M. F. A., Lucas, J., & Yusaf, T. (2022). Effects of biofuel on engines performance and emission characteristics: A review. *Energy*, 238, 121910. https://doi.org/10.1016/j.energy.2021.121910.

Altarazi, Y. S. M., Yusaf, T., Yu, J., Gires, E., Ghafir, M. F. A., & Lucas, J. (2022). A review of engine performance and emissions using single and dual biodiesel fuels: Research paths, challenges, motivations and recommendations. *Fuel*, 326, 125072. https://doi.org/10.1016/j.fuel.2022.125072.

Anand, K., Sharma, R. P., & Mehta, P. S. (2011). A comprehensive approach for estimating thermo-physical properties of biodiesel fuels. *Applied Thermal Engineering*, 31(2–3), 235–242. https://doi.org/10.1016/j.applthermaleng.2010.09.003.

Atadashi, I. M., Aroua, M. K., Aziz, A. R. A., & Sulaiman, N. M. N. (2012). Production of biodiesel using high free fatty acid feedstocks. *Renewable and Sustainable Energy Reviews,* 16(5), 3275–3285. https://doi.org/10.1016/j.rser.2012.02.063.

Athew, G. M., Raina, D., Narisetty, V., Kumar, V., Saran, S., Pugazhendi, A., Sindhu, R., Pandey, A., & Binod, P. (2021). Recent advances in biodiesel production: Challenges and solutions. *Science of the Total Environment*, 794, 148751.

Baukal, C. E., Hayes, R., Grant, M., Singh, P., & Foote, D. (2004). Nitrogen oxides emissions reduction technologies in the petrochemical and refining industries. *Environmental Progress*, 23(1), 19–28.

Baukal Jr, C. E. (2010). *Industrial combustion testing*. CRC Press.

Bazooyar, B., Hallajbashi, N., Shariati, A., & Ghorbani, A. (2014). An investigation of the effect of input air upon combustion performance and emissions of biodiesel and diesel fuel in an experimental boiler. *Energy Sources, Part A: Recovery, Utilization, and Environmental Effects*, 36(4), 383–392. https://doi.org/10.1080/15567036.2010.538810.

Berger, L., Attili, A., & Pitsch, H. (2022). Intrinsic instabilities in premixed hydrogen flames: parametric variation of pressure, equivalence ratio, and temperature. Part 2– Non-linear regime and flame speed enhancement. *Combustion and Flame*, 240, 111936. https://doi.org/10.1016/j.combustflame.2021.111936.

Chen, Z., Chen, H., Wang, L., Geng, L., & Zeng, K. (2020). Parametric study on effects of excess air/fuel ratio, spark timing, and methanol injection timing on combustion characteristics and performance of natural gas/methanol dual-fuel engine at low loads. *Energy Conversion and Management*, 210, 112742. https://doi.org/10.1016/j.enconman.2020.112742.

Chen, Z., Wang, Q., Zhang, X., Zeng, L., Zhang, X., He, T., Liu, T., & Li, Z. (2017). Industrial-scale investigations of anthracite combustion characteristics and NOx emissions in a retrofitted 300 MWe down-fired utility boiler with swirl burners. *Applied Energy*, 202, 169–177. https://doi.org/10.1016/j.apenergy.2017.05.138.

Chu, H., Xiang, L., Meng, S., Dong, W., Gu, M., & Li, Z. (2021). Effects of N2 dilution on laminar burning velocity, combustion characteristics and NOx emissions of rich CH4–air premixed flames. *Fuel*, 284, 119017. https://doi.org/10.1016/j.fuel.2020.119017.

Ciupek, B., Urbaniak, R., Kinalska, D., & Nadolny, Z. (2024). Flue Gas Recirculation System for Biomass Heating Boilers—Research and Technical Applications for Reductions in Nitrogen Oxides (NOx) Emissions. *Energies*, 17(1), 259. https://doi.org/10.3390/en17010259.

Daho, T., Vaitilingom, G., Sanogo, O., Ouiminga, S. K., Zongo, A. S., Piriou, B., & Koulidiati, J. (2014). Combustion of vegetable oils under optimized conditions of atomization and granulometry in a modified fuel oil burner. *Fuel*, 118, 329–334. https://doi.org/10.1016/j.fuel.2013.11.009.

Dey, S., Reang, N. M., Das, P. K., & Deb, M. (2021). A comprehensive study on prospects of economy, environment, and efficiency of palm oil biodiesel as a renewable fuel. *Journal of Cleaner Production*, 286, 124981. https://doi.org/10.1016/j.jclepro.2020.124981.

El-Shenawy, E. A., ElKelawy, M., Bastawissi, H. A.-E., & Al-Hofy, M. (2018). Investigation of Combustion and Emission Characteristics of Diesel/Bioï¿ ¾Diesel Blends under Oxygenated Fuel Burner. *Engineering Research Journal*, 41(2), 109–118. https://doi.org/10.21608/erjm.2018.66297.

Elkelawy, M., Bastawissi, H. A.-E., Abdel-Rahman, A. K., Abou El-Yazied, A., & Mostafa El malla, S. (2023). Effect of multifunctional fuel additive in diesel/waste oil biodiesel blends on industrial burner flame performance and emission characteristics. *International Journal of Ambient Energy*, 44(1), 1382–1395. https://doi.org/10.1080/01430750.2023.2173652.

Ershov, M. A., Savelenko, V. D., Makhova, U. A., Makhmudova, A. E., Zuikov, A. V, Kapustin, V. M., Abdellatief, T. M. M., Burov, N. O., Geng, T., & Abdelkareem, M. A. (2023). Current challenge and innovative progress for producing HVO and FAME biodiesel fuels and their applications. *Waste and Biomass Valorization*, 14(2), 505–521. https://doi.org/10.1007/s12649-022-01880-0.

Esakki, T., Rangaswamy, S. M., & Jayabal, R. (2022). An experimental study on biodiesel production and impact of EGR in a CRDI diesel engine propelled with leather industry waste fat biodiesel. *Fuel*, 321, 123995. https://doi.org/10.1016/j.fuel.2022.123995.

Fathurrahman, N. A., Wibowo, C. S., Bethari, S. A., Anggarani, R., & Aisyah, L. (2021). Fuel properties of two types high-speed diesel blending with palm oil biodiesel in Indonesia. *IOP Conference Series: Earth and Environmental Science*, 749(1), 12030. https://doi.org/10.1088/1755-1315/749/1/012030.

Gad, M. S., Mahfouz, A., & Emara, A. (2021). Spray and combustion characteristics for light diesel/waste cooking oils blended with fuel additives inside an industrial boiler. *Fuel*, 286, 119247. https://doi.org/10.1016/j.fuel.2020.119247.

Gaur, R. K., & Goyal, R. (2022). A review: Effect on performance and emission characteristics of waste cooking oil Biodiesel-diesel blends on IC engine. *Materials Today: Proceedings,* 63, 643–646. https://doi.org/10.1016/j.matpr.2022.04.447.

Ghorbani, A., Bazooyar, B., Shariati, A., Jokar, S. M., Ajami, H., & Naderi, A. (2011). A comparative study of combustion performance and emission of biodiesel blends and diesel in an experimental boiler. *Applied Energy*, 88(12), 4725–4732. https://doi.org/10.1016/j.apenergy.2011.06.016.

Hadi, K., Ichimura, R., Hashimoto, G., Xia, Y., Hashimoto, N., & Fujita, O. (2021). Effect of fuel ratio of coal on the turbulent flame speed of ammonia/coal particle cloud co-combustion at atmospheric pressure. *Proceedings of the Combustion Institute*, 38(3), 4131–4139.

Hashimoto, N., Nishida, H., & Ozawa, Y. (2014). Fundamental combustion characteristics of Jatropha oil as alternative fuel for gas turbines. *Fuel*, 126, 194–201. https://doi.org/10.1016/j.fuel.2014.02.057.

Hosseini, S. B., Bashirnezhad, K., Moghiman, A. R., Khazraii, Y., & Nikoofal, N. (2010). Experimental comparison of combustion characteristic and pollutant emission of gas oil and biodiesel. International Scholarly and Scientific Research & Innovation 4(12) 2010.

Huang, H., Tian, J., Li, J., & Tan, D. (2022). Effects of different exhaust gas recirculation (EGR) rates on combustion and emission characteristics of biodiesel–diesel blended fuel based on an improved chemical mechanism. *Energies*, 15(11), 4153. https://doi.org/10.3390/en15114153.

Kamaraj, R., & Rao, Y. K. S. S. (2022). Biodiesel blends: a comprehensive systematic review on various constraints. *Environmental Science and Pollution Research*, 1–16.

Kass, M., Kaul, B., Armstrong, B., Szybist, J., & Lobodin, V. (2022). Stability, rheological and combustion properties of biodiesel blends with a very low sulfur fuel oil (VLSFO). *Fuel*, 316, 123365. https://doi.org/10.1016/j.fuel.2022.123365.

Kaya, C., & Kökkülünk, G. (2023). Biodiesel as alternative additive fuel for diesel engines: an experimental and theoretical investigation on emissions and performance characteristics. *Energy Sources, Part A: Recovery, Utilization, and Environmental Effects,* 45(4), 10741–10763. https://doi.org/10.1080/15567036.2020.1774685.

Khalid, A., Lambosi, L., Lokman, M. M., Manshoor, B., Zaman, I. Bin, Sapit, A., & Amirnordin, S. H. (2014). Effect of Biodiesel from Waste Cooking Oil on Mixture Formation and Emission of Burner Combustion. *Applied Mechanics and Materials*, 607, 620–624. https://doi.org/10.4028/www.scientific.net/AMM.607.620.

Khodir, A. K., Asar, G. M., El-Behery, S. M., & El-Askary, W. A. (2019). Effect of atomizing air swirl angle on combustion and emission characteristics of Spray Flame. *IOSR Journal of Mechanical and Civil Engineering*, 16(2), 1–11.

Kikuchi, K., Hori, T., & Akamatsu, F. (2022). Fundamental study on hydrogen low-NOx combustion using exhaust gas self-recirculation. *Processes*, 10(1), 130. https://doi.org/10.3390/pr10010130.

Kim, H. J., Jo, S., Kwon, S., Lee, J.-T., & Park, S. (2022). NOX emission analysis according to after-treatment devices (SCR, LNT+ SCR, SDPF) and control strategies in Euro-6 light-duty diesel vehicles. *Fuel*, 310, 122297. https://doi.org/10.1016/j.fuel.2021.122297.

Kim, Y., Park, C., Oh, J., Oh, S., Choi, Y., & Lee, J. (2023). Effect of excessive air ratio on hydrogen-fueled spark ignition engine with high compression ratio using direct injection system toward higher brake power and thermal efficiency. *International Journal of Automotive Technology*, 24(1), 79–89. https://doi.org/10.1007/s12239-023-0008-7.

Kulkarni, S. J., Suryawanshi, M. A., Mane, V. B., & Kumbhar, G. (2023). Biohydrogen From Waste Feedstocks–Materials, Methods and Recent Developments. B*ioNanoScience*, 13(4), 1501–1516. https://doi.org/10.1007/s12668-023-01206-x.

Kumaran, P., Gopinathan, M., & Kantharrajan, S. (2014). Combustion characteristics of improved biodiesel in diffusion burner. I*nternational Journal of Automotive and Mechanical Engineering*, 10, 2112. https://doi.org/10.15282/ijame.10.2014.27.0178.

Kwon, M., Nguyen, B. H., Kim, S., Kim, Y., & Park, J. H. (2018). Numerical investigation of buoyancy and thermal radiation effects on a mid-/large-sized low NOx combustion system with flue-gas internal recirculation. *Advances in Mechanical Engineering,* 10(4), 1687814018769139. https://doi.org/10.1177/1687814018769139.

Li, J., Zhang, X., Yang, W., & Blasiak, W. (2013). Effects of flue gas internal recirculation on NOx and SOx emissions in a co-firing boiler. *International Journal of Clean Coal and Energy*, 2(2), 13–21. https://doi.org/10.4236/ijcce.2013.22002.

Linganiso, E. C., Tlhaole, B., Magagula, L. P., Dziike, S., Linganiso, L. Z., Motaung, T. E., Moloto, N., & Tetana, Z. N. (2022). Biodiesel production from waste oils: a South African outlook. *Sustainability*, 14(4), 1983. https://doi.org/10.3390/su14041983.

Liu, C., Lai, T., Zhang, H., Huang, P., Lin, K., Lin, Y., & Li, M. (2024). Experimental study on the morphology of turbulent diffusion flames with different aspect ratios under slope conditions. *Fuel*, 357, 129827. https://doi.org/10.1016/j.fuel.2023.129827.

Macor, A., & Pavanello, P. (2009). Performance and emissions of biodiesel in a boiler for residential heating. *Energy*, 34(12), 2025–2032. https://doi.org/10.1016/j.energy.2008.08.021.

Mahfouz, A., Gad, M. S., El Fatih, A., & Emara, A. (2018). Comparative study of combustion characteristics and exhaust emissions of waste cooking-diesel oil blends. *Ain Shams Engineering Journal*, 9(4), 3123–3134. https://doi.org/10.1016/j.asej.2018.03.004.

Manikandan, G., Kanna, P. R., Taler, D., & Sobota, T. (2023). Effects of flue gas internal recirculation on NOx and SOx emissions in a co-firing boiler. Energies, 16(4), 1739.

Masera, K., & Hossain, A. K. (2023). Advancement of biodiesel fuel quality and NOx emission control techniques. *Renewable and Sustainable Energy Reviews*, 178, 113235. https://doi.org/10.1016/j.rser.2023.113235.

Milano, J., Shamsuddin, A. H., Silitonga, A. S., Sebayang, A. H., Siregar, M. A., Masjuki, H. H., Pulungan, M. A., Chia, S. R., & Zamri, M. (2022). Tribological study on the biodiesel produced from waste cooking oil, waste cooking oil blend with Calophyllum inophyllum and its diesel blends on lubricant oil. *Energy Reports*, 8, 1578–1590. https://doi.org/10.1016/j.egyr.2021.12.059.

Misra, P., Takigawa, M., Khatri, P., Dhaka, S. K., Dimri, A. P., Yamaji, K., Kajino, M., Takeuchi, W., Imasu, R., & Nitta, K. (2021). Nitrogen oxides concentration and emission change detection during COVID-19 restrictions in North India. *Scientific Reports,* 11(1), 9800. https://doi.org/10.1038/s41598-021-87673-2.

Mizik, T., & Gyarmati, G. (2021). Economic and sustainability of biodiesel production—a systematic literature review. *Clean Technologies*, 3(1), 19–36. https://doi.org/10.3390/cleantechnol3010002.

Mourad, M., Mahmoud, K. R. M., & NourEldeen, E.-S. H. (2021). Improving diesel engine performance and emissions characteristics fueled with biodiesel. *Fuel*, 302, 121097. https://doi.org/10.1016/j.fuel.2021.121097.

Mousavi, H., Mortazavi, Y., Khodadadi, A. A., Saberi, M. H., & Alirezaei, S. (2021). Enormous enhancement of Pt/SnO2 sensors response and selectivity by their reduction, to CO in automotive exhaust gas pollutants including CO, NOx and C3H8. *Applied Surface Science*, 546, 149120. https://doi.org/10.1016/j.apsusc.2021.149120.

Ng, H. K., & Gan, S. (2010). Combustion performance and exhaust emissions from the non-pressurised combustion of palm oil biodiesel blends. *Applied Thermal Engineering*, 30(16), 2476–2484. https://doi.org/10.1016/j.applthermaleng.2010.06.020.

Nhan, H. K., Kwon, M., Kim, S., & Park, J. H. (2019). CFD investigation of NO x reduction with a flue-gas internal recirculation burner in a mid-sized boiler. *Journal of Mechanical Science and Technology*, 33, 2967–2978. https://doi.org/10.1007/s12206-019-0546-0.

Norwazan, A. R., Jaafar, M. N. M., Sapee, S., & Farouk, H. (2018). Emissions of Jatropha oil-derived biodiesel blend fuels during combustion in a swirl burner. *IOP Conference Series: Earth and Environmental Science,* 136(1), 12001. https://doi.org/10.1088/1755-1315/136/1/012001.

Palani, Y., Devarajan, C., Manickam, D., & Thanikodi, S. (2022). Performance and emission characteristics of biodiesel-blend in diesel engine: A review. *Environmental Engineering Research*, 27(1). https://doi.org/10.4491/eer.2020.338.

Pan, X. (2021). *Biodiesel instead of fossil fuels in diesel vehicles*. Thesis. Centria University of Applied Sciences.

Pandey, K. K., & Murugan, S. (2022). A review of bio-fuelled LHR engines. *International Journal of Ambient Energy*, 43(1), 2486–2509. https://doi.org/10.1080/01430750.2020.1712254.

Patil, A. R., Kakati, D., Singh, B., Rosen, M. A., Patil, R., Javanjal, V., Sonawane, C., Panchal, H., Kumar, A., & Siddiqui, M. I. H. (2024). Experimental investigation of multi-additive fuel blend and its optimization for CI engine performance and emissions by the hybrid Taguchi-TOPSIS technique. *Case Studies in Thermal Engineering*, 53, 103703. https://doi.org/10.1016/j.csite.2023.103703.

Pei, X., Abdul Jameel, A. G., Chen, C., AlGhamdi, I. A., AlAhmadi, K., AlBarakati, E., Saxena, S., & Roberts, W. L. (2021). Swirling flame combustion of heavy fuel oil: Effect of fuel sulfur content. *Journal of Energy Resources Technology*, 143(8), 82103. https://doi.org/10.1115/1.4048942.

Pourhoseini, S. H., & Ghodrat, M. (2021). Experimental investigation of the effect of Al2O3 nanoparticles as additives to B20 blended biodiesel fuel: Flame characteristics, thermal performance and pollutant emissions. *Case Studies in Thermal Engineering*, 27, 101292. https://doi.org/10.1016/j.csite.2021.101292.

Pourhoseini, S. H., Taghvaei, I., Moghiman, M., & Baghban, M. (2021). Tangential Flue Gas Recirculation (TFGR) technique for enhancement of radiation characteristics and reduction of NOx emission in natural gas burners. *Journal of Natural Gas Science and Engineering*, 94, 104130. https://doi.org/10.1016/j.jngse.2021.104130.

Prasad, S., Kumar, S., Sheetal, K. R., & Venkatramanan, V. (2020). Global climate change and biofuels policy: Indian perspectives. *Global Climate Change and Environmental Policy: Agriculture Perspectives*, 207–226.

Razzaq, L., Imran, S., Anwar, Z., Farooq, M., Abbas, M. M., Mehmood Khan, H., Asif, T., Amjad, M., Soudagar, M. E. M., & Shaukat, N. (2020). Maximising yield and engine efficiency using optimised waste cooking oil biodiesel. *Energies*, 13(22), 5941.

Rehman, A., Phalke, D. R., & Pandey, R. (2011). Alternative fuel for gas turbine: Esterified jatropha oil–diesel blend. *Renewable Energy*, 36(10), 2635–2640. https://doi.org/10.1016/j.renene.2010.06.013.

Samuel, O. D., Okwu, M. O., Oyejide, O. J., Taghinezhad, E., Afzal, A., & Kaveh, M. (2020). Optimizing biodiesel production from abundant waste oils through empirical method and grey wolf optimizer. *Fuel*, 281, 118701. https://doi.org/10.1016/j.fuel.2020.118701.

Samuel, O. D., Waheed, M. A., Taheri-Garavand, A., Verma, T. N., Dairo, O. U., Bolaji, B. O., & Afzal, A. (2021). Prandtl number of optimum biodiesel from food industrial waste oil and diesel fuel blend for diesel engine. *Fuel*, 285, 119049. https://doi.org/10.1016/j.fuel.2020.119049.

Saraçoğlu, N., & Gündüz, G. (2009). *Energy sources, part a: recovery, utilization, and environmental effects*. Report, Bartin University Institutional Repository http://hdl.handle.net/11772/1779.

Sarwer, A., Hussain, M., Al-Muhtaseb, A. H., Inayat, A., Rafiq, S., Khurram, M. S., Ul-Haq, N., Shah, N. S., Alaud Din, A., & Ahmad, I. (2022). Suitability of biofuels production on commercial scale from various feedstocks: a critical review. *ChemBioEng Reviews*, 9(5), 423–441. https://doi.org/10.1002/cben.202100049.

Sentanuhady, J., Hasan, W. H., & Muflikhun, M. A. (2022). Recent progress on the implementation of renewable biodiesel fuel for automotive and power plants: raw materials perspective. *Advances in Materials Science and Engineering*, 2022(1), 5452942. https://doi.org/10.1155/2022/5452942.

Sharma, Y. C., & Singh, B. (2008). Development of biodiesel from karanja, a tree found in rural India. Fuel, 87(8–9), 1740–1742. https://doi.org/10.1016/j.fuel.2007.08.001.

Sharon, H., Karuppasamy, K., Kumar, D. R. S., & Sundaresan, A. (2012). A test on DI diesel engine fueled with methyl esters of used palm oil. *Renewable Energy*, 47, 160–166. https://doi.org/10.1016/j.renene.2012.04.032.

Simsek, S. (2020). Effects of biodiesel obtained from Canola, sefflower oils and waste oils on the engine performance and exhaust emissions. *Fuel*, 265, 117026. https://doi.org/10.1016/j.fuel.2020.117026.

Singh, S. P., & Singh, D. (2010). Biodiesel production through the use of different sources and characterization of oils and their esters as the substitute of diesel: a review. *Renewable and Sustainable Energy Reviews*, 14(1), 200–216. https://doi.org/10.1016/j.rser.2009.07.017.

Suardi, M., Amirnordin, S. H., & Khalid, A. (2006). *Effect of mixture formation on combustion in burner system.*

Subbaramaiah, K., & Sujatha, P. (2023). Optimal DG unit placement in distribution networks by multi-objective whale optimization algorithm & its techno-economic analysis. *Electric Power Systems Research*, 214, 108869. https://doi.org/10.1016/j.epsr.2022.108869.

Suhara, A., Karyadi, Herawan, S. G., Tirta, A., Idris, M., Roslan, M. F., Putra, N. R., Hananto, A. L., & Veza, I. (2024). Biodiesel Sustainability: Review of Progress and Challenges of Biodiesel as Sustainable Biofuel. *Clean Technologies*, 6(3), 886–906. https://doi.org/10.3390/cleantechnol6030045.

Sun, C., Liu, Y., Qiao, X., Ju, D., Tang, Q., Fang, X., & Zhou, F. (2020). Experimental study of effects of exhaust gas recirculation on combustion, performance, and emissions of DME-biodiesel fueled engine. *Energy*, 197, 117233. https://doi.org/10.1016/j.energy.2020.117233.

Suzihaque, M. U. H., Alwi, H., Ibrahim, U. K., Abdullah, S., & Haron, N. (2022). Biodiesel production from waste cooking oil: A brief review. *Materials Today: Proceedings*, 63, S490–S495.

Tan, K. T., Lee, K. T., & Mohamed, A. R. (2011). Potential of waste palm cooking oil for catalyst-free biodiesel production. *Energy*, 36(4), 2085–2088. https://doi.org/10.1016/j.energy.2010.05.003.

Tanasić, N., Stamenić, M., & Tanasić, V. (2016). S. H. Pourhoseini, I. Taghvaei, M. N. Tanasić, M. Stamenić, V. Tanasić, Effects of Flue Gas Recirculation on NOx Emissions from Gas-Fired Utility Boilers, in: Curr. Probl. Exp. Comput. Eng. Proc. Int. Conf. Exp. Numer. Investig. New Technol gas burners. *Journal of Al-Azhar University Engineering Sector*, 11(41), 1265–1274.

Tanasić, N., Stamenić, M., & Tanasić, V. (2022). Effects of Flue Gas Recirculation on NOx Emissions from Gas-Fired Utility Boilers. Current Problems in Experimental and Computational Engineering: *Proceedings of the International Conference of Experimental and Numerical Investigations and New Technologies*, CNNTech 2021, 319–337. https://doi.org/10.1007/978-3-030-86009-7_17.

Tang, G., Liu, Y., Zhang, J., Liu, B., Li, Q., Sun, J., Wang, Y., Xuan, Y., Li, Y., & Pan, J. (2021). Bypassing the NOx titration trap in ozone pollution control in Beijing. *Atmospheric Research*, 249, 105333. https://doi.org/10.1016/j.atmosres.2020.105333.

Wang, H., Ji, C., Su, T., Shi, C., Ge, Y., Yang, J., & Wang, S. (2022). Comparison and implementation of machine learning models for predicting the combustion phases of hydrogen-enriched Wankel rotary engines. *Fuel*, 310, 122371. https://doi.org/10.1016/j.fuel.2021.122371.

Wang, Q., Chen, Z., Wang, L., Zeng, L., & Li, Z. (2018). Application of eccentric-swirl-secondary-air combustion technology for high-efficiency and low-NOx performance on a large-scale down-fired boiler with swirl burners. *Applied Energy*, 223, 358–368. https://doi.org/10.1016/j.apenergy.2018.04.064.

Xia, Y., Hashimoto, G., Hadi, K., Hashimoto, N., Hayakawa, A., Kobayashi, H., & Fujita, O. (2020). Turbulent burning velocity of ammonia/oxygen/nitrogen premixed flame in O2-enriched air condition. *Fuel*, 268, 117383. https://doi.org/10.1016/j.fuel.2020.117383.

Yesilyurt, M. K., & Aydin, M. (2020). Experimental investigation on the performance, combustion and exhaust emission characteristics of a compression-ignition engine fueled with cottonseed oil biodiesel/diethyl ether/diesel fuel blends. *Energy Conversion and Management*, 205, 112355. https://doi.org/10.1016/j.enconman.2019.112355.

Yesilyurt, M. K., Aydin, M., Yilbasi, Z., & Arslan, M. (2020). Investigation on the structural effects of the addition of alcohols having various chain lengths into the vegetable oil-biodiesel-diesel fuel blends: An attempt for improving the performance, combustion, and exhaust emission characteristics of a compression ignition engine. *Fuel*, 269, 117455. https://doi.org/10.1016/j.fuel.2020.117455.

Yilmaz, N., Atmanli, A., Hall, M. J., & Vigil, F. M. (2022). Determination of the optimum blend ratio of diesel, waste oil derived biodiesel and 1-pentanol using the response surface method. *Energies*, 15(14), 5144. https://doi.org/10.3390/en15145144.

Yip, A., Haelssig, J. B., & Pegg, M. J. (2021). Multicomponent pool fires: Trends in burning rate, flame height, and flame temperature. *Fuel*, 284, 118913. https://doi.org/10.1016/j.fuel.2020.118913.

Yusoff, M. N. A. M., Zulkifli, N. W. M., Sukiman, N. L., Chyuan, O. H., Hassan, M. H., Hasnul, M. H., Zulkifli, M. S. A., Abbas, M. M., & Zakaria, M. Z. (2021). Sustainability of palm biodiesel in transportation: a review on biofuel standard, policy and international collaboration between Malaysia and Colombia. *Bioenergy Research*, 14(1), 43–60. https://doi.org/10.1007/s12155-020-10165-0.

Zare, A., Stevanovic, S., Jafari, M., Verma, P., Babaie, M., Yang, L., Rahman, M. M., Ristovski, Z. D., Brown, R. J., & Bodisco, T. A. (2021). Analysis of cold-start NO2 and NOx emissions, and the NO_2/NOx ratio in a diesel engine powered with different diesel-biodiesel blends. *Environmental Pollution*, 290, 118052. https://doi.org/10.1016/j.envpol.2021.118052.

Zhang, R., Zhang, Y., Lin, H., Feng, X., Fu, T.-M., & Wang, Y. (2020). NOx emission reduction and recovery during COVID-19 in East China. *Atmosphere*, 11(4), 433. https://doi.org/10.3390/atmos11040433.

Zhang, Z., Zhou, L., He, X., Chen, L., & Wei, H. (2023). Kumaran, P., Gopinathan, M., & Kantharrajan, S. (2014). Combustion characteristics of improved biodiesel in diffusion burner. *International Journal of Automotive and Mechanical Engineering*, 10, 2112. Combustion and Flame, 251, 112719. https://doi.org/10.1016/j.combustflame.2023.112719.

Zhu, Y., Wang, C., & Chen, X. (2021). Combustion characteristic study with a flue gas internal and external double recirculation burner. *Chemical Engineering and Processing-Process Intensification*, 162, 108345. https://doi.org/10.1016/j.cep.2021.108345.

Chapter 6

An Outlook on the Geospatial Analysis Approach to Employ Functional Indicators for Regional Sustainable Development Goals

Doaa Taha Yaseen[1,2,*]
and Oday Zakariya Jasim[2]
[1]Department of Civil Engineering, Wasit University, Wasit, Iraq
[2]Department of Civil Engineering, University of Technology, Baghdad, Iraq

Abstract

This review examines the sustainable development goals from a regional perspective to determine their success and draw attention to the problems they face. Understanding the necessary spatial patterns is crucial for achieving the sustainable development goals. Dealing with world environmental issues becomes more important, this is especially true. This study was performed to find out how well the Sustainable Development Goals (SDGs) have been met around the world using the newest tools for regional research and analysis. The SDGs cover a wide range of issues, including decreasing poverty, boosting the economy, and making health care and schooling better. Research in this area uses remote sensing, Geographic Information Systems (GIS), and spatial models to consider how geography affects important long-term goals for sustainable growth in different regions. The results look at the good and bad aspects of implementing the SDGs and provide insight into the achievements of different projects. Sustainable development success

[*] Corresponding Author's Email: bce.22.13@grad.uotechnology.edu.iq

In: Advancements in Sustainability Systems II
Editors: H. H. Al-Kayiem, R. A. Jessam, Hasanain A. A. Wahhab et al.
ISBN: 979-8-89530-562-1

rates can help you find trouble spots and better distribute resources. Showing how the Sustainable Development Goals (SDGs) are linked in the study suggests that sector-level collaboration and coherent policies are very important. According to the researchers, sustainable development goals could be achieved faster if politicians focused on specific projects and used data to help them make decisions. Looking at sustainable development from a geographic perspective adds to the current conversation about it. As a result, it is easy to see the various difficulties and opportunities for achieving the SDGs.

Keywords: geospatial analysis, geospatial data and tools, indicators, sustainable development goals (SDGs)

1. Introduction

Geospatial analysis is one of the most important tools in data science, and it is also an important part of the journey to reach sustainable development goals. Hunger, poverty, health, education, climate change, equal rights between men and women, energy, water and sanitation, transport, social justice, and other issues are covered by the sustainable development goals. The sustainable development goals are a big part of what the UN wants to do after 2015 Oluboyede & Odeyemi, 2020. Inspired by the Millennium Development Goals, the 2030 Agenda includes 17 sustainable development goals aimed at making the world richer and more sustainable Avtar et al. 2020.

Geospatial analysis technology can be used to look at data that has a real-world component. This is generally done using GIS. Using location data to learn more about land use and other parts of urban growth is one example of Jasim et al. 2019. Shared, low-cost data sources, such as images collected through remote sensing, can help achieve sustainable development goals and maintain their relevance at the global level. Using spatial data analysis is essential to tracking things, making choices, and doing accounting. Many people said that this study is important for understanding how to evaluate the sustainable development goals of Das et al. 2023. Therefore, without satellite images, it is impossible to create or update location data and maps (Nashait et al. 2020).

These opening paves the way for considering how geographical research and SDGs can work together. This study shows how important it is to look at problems in terms of space when trying to understand and solve them. It talks about geographical research, the objectives of the study, and its structure. This study aims to demonstrate how geographic analysis can help achieve the SDGs and add to what is already known in the field by considering important methods, uses, problems, and future possibilities.

Then the paper moves to the next part: Section (2) explains the methods mixed with the previous study. The third part is where we talk about the sustainable development goals. In Section (4), we talk about regional research's role in promoting sustainable development. The fifth section deals with geospatial techniques and case studies in the field of SDGs. Sections (6) and (7) cover the results and conclusions.

2. Methodology

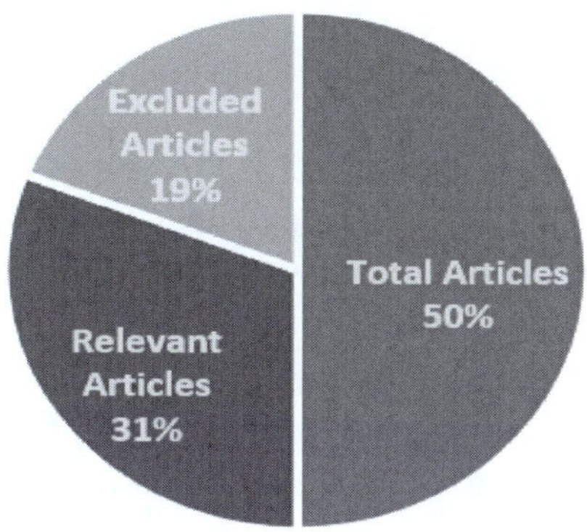

Source: Authors.

Figure 1. Literature collecting and filtering process.

In this study, a systematic review method was used, aiming to learn more about how to put the SDGs into practice. The systematic review and analysis method involved a careful and systematic search for and access to relevant written materials that spoke to putting the SDGs into practice. To accomplish the target of the study, the main databases and previous contribution sources, such as Google Scholar, Scopus, Web of Knowledge, and Science Direct, have been searched. The main searched keywords are "sustainable development goals," "geospatial technology and the sustainable development goals," "geospatial data and the SDGs," "remote sensing, GIS and the SDGs," and "monitoring progress on the SDGs." There were no time limits to the research, and the focus was on finding information that added much to the current conversation about the SDGs. However, more current papers have been collected to show that the topic is becoming more important. The search terms to be used and how to include or exclude stories were determined. Articles unrelated to the SDGs were excluded from the study, but articles relevant to the topic were added. The selected papers have been classified and clustered to capture the most important ideas and findings, considering their relevance to the topic and how well they are written. The paper will conclude by analyzing the major themes and patterns discovered in the literature and their implications for future research. Figure 1 shows a literature collecting and filtering process in the current attempt to develop an outlook on the geospatial approaches to employ functional indicators for regional SDGs.

3. Sustainable Development Goals

The SDGs, set by the United Nations, are globally accepted goals to enhance the well-being of billions of people in developed, emerging, and developing countries. They were adopted in 2015 as part of the 2030 Agenda for Sustainable Development, a global commitment to end poverty, protect the planet, and ensure prosperity for all by 2030 (Arora & Mishra, 2019). The 17 Sustainable Development Goals address a broad range of global challenges, including poverty, inequality, climate change, environmental degradation, peace, and justice. They focus on saving the environment, maintaining a healthy economy, and improving social justice (Arora & Mishra, 2019). To track and evaluate progress toward these goals, the United Nations has developed a set of 232 indicators. These indicators, known as the UN SDG Indicators, are used to monitor, measure, and evaluate the implementation of the SDGs worldwide. Each indicator corresponds to a specific goal within one of the 17 goals, ensuring that each aspect of sustainable development is appropriately monitored. The indicators are classified into three groups based on the development of their methodologies and the availability of data for global monitoring (Nicolau et al. 2018).

4. Roles of Geospatial Analysis in Sustainable Development

Geospatial analysis contributes significantly to sustainable development by giving important insights into a variety of environmental, social, and economic aspects. Geospatial technologies use a number of monitoring devices to collect data and monitor projects that promote sustainable development. Remote sensing data is frequently accessible and readily downloadable (Dhumad et al. 2023). Satellite devices, through remote sensing methods, collect images that constitute a crucial source of data for tracking and monitoring natural and man-made features of the Earth (Khalaf et al. 2018). Air temperature can be found using weather stations, remote sensing, and climate reanalysis datasets Jasim et al. 2020. The combination of RS data, in situ tracking, and precise geographic information allows natural resources to be measured and comprehensive hazard and risk assessments to be completed. This data is very useful for making policies, plans, and running projects to achieve sustainable growth (Acharya & Lee, 2019). Social and business actions, as well as the environment, are all connected. Without looking at the aspect of space, it is impossible to understand and see the changes happening. GIS gives many options for working with data in location files Jasim et al. 2018. GISs deal with two types of data (geographic and non-spatial) (Hashim et al. 2021). New information and technology fields, like machine learning and artificial intelligence, can be used together to get more accurate estimates, predictions, and studies of a wide range of natural, social, environmental, and other events (Saad et al. 2023).

In terms of making progress toward sustainable development, the 2030 Agenda is a major step forward in how society manages social equality, economic growth, and environmental sustainability. The Sustainable Development Goals (SDGs) and their accompanying global indicator architecture create a data-driven structure that countries can use to make decisions and build fact-based policies. The Agenda knows that "if you cannot measure it, you cannot manage it," and data is what makes the 2030 Agenda possible., the UN knows that place is important for bringing together data about society, the economy, and the environment, as well as for keeping track of all the SDGs (Oluboyede & Odeyemi, 2020). Figure 2 illustrates the schematic diagram of the framework for geospatial analysis in achieving sustainable development.

Geomatics is the study of the Earth's surface and its surrounding areas with the help of instruments. Remote sensors, GIS tools, and a global signature system are used. Using computers, it can connect digital technology, space, and public images. This makes it easier to find information that can help make choices (Husein et al. 2018). To obtain an accurate position in aerial mapping, ground control points (GCPs) are necessary (Hilal et al. 2021). Viewing the geographic coordinates of ground control points is essential to obtaining accurate locations, and each of these ground control points must have a small error for readings to be extremely accurate (Hilal et al. 2022).

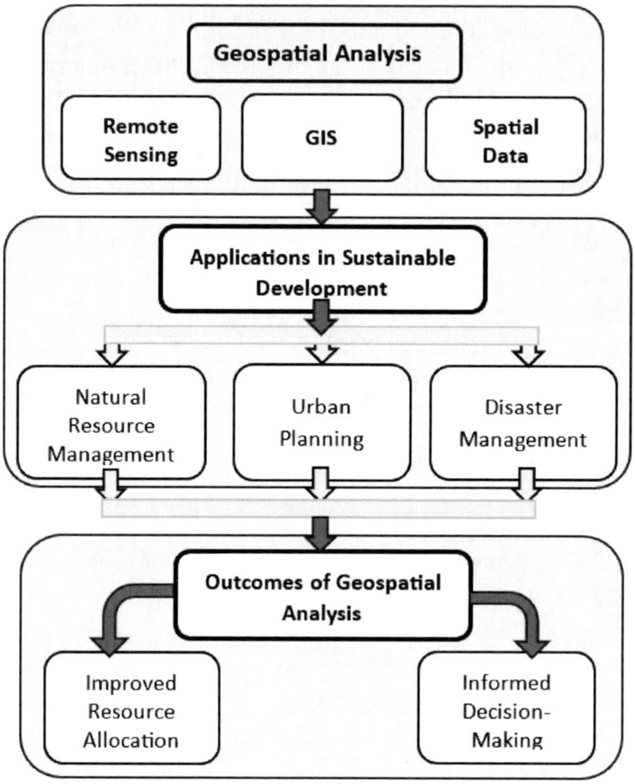

Source: Authors.

Figure 2. Schematic diagram of the framework for geospatial analysis in achieving sustainable development.

Geospatial methods, such as Geographic Knowledge Systems (GIS), have recently shown that they are good at handling information about the Earth Hashim et al. 2022. GIS can look at the real world by adding information on top of each other and integrating it with its location. It can provide a complete picture of the group at many levels of time and space Sadiqe et al. 2023. Interpreting and studying the interactions, connections, and relationships between GIS data types specific to a place is part of the GIS spatial analysis modeling process Karim et al. 2016.

5. Geospatial, Economic, Social, and Environmental Analysis of SDGs

The following parts talk about the literature review and the SDGs, which are closely linked to the economic, social, and environmental aspects of the goals. They also talk about how important geospatial analysis is for sustainable development.

5.1. SDG 1: No Poverty

SDG 1 is to end all forms of poverty worldwide. Wetter et al. 2017 poverty in low- and middle-income countries can be modeled using public and private data sources, such as cell phone company data. The data show that it is possible to closely monitor poverty rates in a detailed and consistent manner. This can provide us with new information in places where resources are limited. Jean et al. 2016 emphasized that the use of the latest machine learning methods and satellite data collected through remote sensing has facilitated poverty mapping and forecasting on a global scale. Elvidge et al. 2009 suggested that high-resolution satellite images could help learn more about the social and economic conditions of people around the world.

5.2. SDG 2: Zero Hunger

SDG 2 aims to eliminate hunger, ensure food security, improve nutrition, and promote sustainable agriculture globally. Shelestov et al. 2017 evaluated the efficiency of using Google Earth Engine (GEE) for classifying multi-temporal satellite imagery, aiming for large-scale crop mapping. It compares various classifiers within GEE and finds that a neural network approach outperforms others in classification accuracy for a large territory in Ukraine. Dell'Acqua et al. 2018 presented a plan that uses worldwide datasets and fusion techniques to create maps of cash crops. These maps will help with insurance and mitigation plans, taking into account limitations and maximizing processing efficiency. Pandey & Pandey, 2023 emphasized that to tackle the issue of global food security, it is necessary to combine agriculture and technology, namely precision agriculture and geospatial technology. This approach should be in line with many Sustainable Development Goals (SDGs). This convergence aims to decrease the number of chemicals used, boost productivity, and protect biodiversity. Geospatial technology is essential for monitoring and improving agricultural practices to promote sustainable growth.

5.3. SDG 3: Good Health and Well-Being

SDG 3 aims to promote universal health and well-being by directing attention towards enhancing healthcare accessibility and diminishing mortality rates. Orimoloye et al. 2018 emphasized the health hazards associated with the rise in land surface temperature and radiation caused by urbanization. It underscores the significance of remote sensing in climate monitoring for health management. Asi & Williams, 2018 utilized digital health technologies can help close healthcare disparities in areas affected by conflict by addressing specific requirements, facilitating education and outreach, enhancing data collection, and promoting joint efforts to achieve Sustainable Development Goal 3 targets.

5.4. SDG 4: Quality Education

Aiming to promote possibilities for lifelong learning and guarantee universal access to quality education, SDG 4 aims to do just that. An analysis of the locations of New York City census tracts shows large groups of high school graduates, racial minorities, and low-income areas. This illustrates the intricate interplay among dropout rates, race, and poverty, thereby influencing the allocation of resources and the implementation of targeted interventions (Trinidad, 2022). The use of data-driven and AI-powered systems can significantly improve basic and secondary education, especially in remote areas with insufficient formal school buildings (Bachmann et al. 2022).

5.5. SDG 5: Gender Equality

Striving for gender equality and giving all women and girls more power is goal 5 of the sustainable development agenda. Artificial intelligence (AI) technologies may help eliminate gender bias in hiring choices (Palomares et al. 2021).

5.6. SDG 6: Clean Water and Sanitation

As part of Sustainable Development Goal 6, which calls for long-term control of the world's water supply, everyone should have access to clean water. Machiwal et al. 2011 demonstrate a method that uses remote sensing, GIS, and multi-criteria decision-making methods to find groundwater-rich places in the Udaipur district of Rajasthan. For good control of groundwater resources, the results of this work are very important. Eguaroje et al. 2020 used the Standardized Precipitation Index (SPI) and NDVI to look at the drought in Kano State, Nigeria, in 2018. The results show that the region had largely normal conditions. Gain et al. 2016 demonstrated a multi-criteria geographic analysis method for assessing water security worldwide using signals from SDG 6. It focuses on parts of Africa and Asia that need comprehensive solutions to meet water-related SDG targets.

5.7. SDG 7: Affordable and Clean Energy

SDG 7 wants to make sure that everyone has access to clean, modern, reliable, and long-lasting energy. Many countries want to use clean energy, such as solar energy, in order to achieve sustainable growth goals (Husein et al. 2023). Hassani et al. 2019 used "big data" technology to look at satellite images and make correct predictions about energy poverty. Based on the results of this study, sub-Saharan Africa has the highest rate of energy poverty.

5.8. SDG 8: Decent Work and Economic Growth

The goals of SDG 8 are full and useful jobs, fair work for everyone, and long-lasting economic growth that benefits everyone. Using data analytics and keeping an eye on signs, big data could help reach the goal of economic growth (target 8.1), Hassani et al. 2021.

5.9. SDG 9: Industry, Innovation, and Infrastructure

SDG 9 wants to build infrastructure that can handle damage, encourage manufacturing that benefits everyone and lasts, and encourage new ideas. Ilie et al. 2019 suggested using global open data and open tools to monitor SDG 9. They offer tools for learning how to use open-source software and high-quality global geospatial information that is easy to find. The training materials offer a systematic and detailed approach. Instructions on how to calculate the status of Sustainable Development Goal (SDG) indicators to facilitate the tracking of progress towards achieving the SDGs. To showcase the practicality of their method, they carried out a case study in the rural region of Tanzania.

5.10. SDG 10: Reduced Inequalities

SDG 10 seeks to minimize disparities within and between nations, with an emphasis on income, gender, age, disability, and other types of discrimination. Truby, 2020 contends that data-driven and AI-driven technologies significantly influence inequality.

5.11. SDG 11: Sustainable Cities and Communities

The incorporation of geospatial techniques in smart cities allows the resolution of urban issues and the encouragement of innovation towards developing more sustainable urban settings, in line with the aims of the SDGs (Costa et al. 2024). Laituri et al. 2021 evaluated the cities of Denpasar, Kharkiv, and Mekelle as components of the Secondary Cities (2C) initiative. It uses field-based collaborative mapping to test how well SDG 11.3.1 works, but the lack of data makes it look like it may not work well in small and medium-sized cities. Brussel et al. 2019 critically evaluated the UN method for measuring city accessibility with reference to SDG metric 11.2. It presents different measures of accessibility that give a fuller picture of transportation systems and the differences that emerge in urban areas. Bonafoni et al. 2017 used Landsat 7 data to look at city roofs and found changes in reflectance and ground surface temperature in Terni, Italy, over 10 years. The study shows that remote sensing can be used to check the resilience of cities and reduce urban heat island effects. Costa et al. 2024 emphasize the importance of using data-driven approaches to encourage long-term smart city growth. It focuses on addressing the complexities of urban dynamics and standardizing geospatial data. It also advocates for the use of creative approaches and research that prioritize sustainable objectives. Kuffer et al. 2018 investigated the efficacy of using earth observation-based mapping to identify impoverished living conditions in cities in developing countries. The goal is to help reach SDG indicator 11.1.1 by figuring out technical and operational problems and suggesting possible additions to a global spatial information database that can be used for targeted interventions.

5.12. SDG 12: Responsible Consumption and Production

SDG 12 seeks to guarantee sustainable patterns of production and consumption, encompassing waste minimization, resource efficiency, and promoting sustainable practices across supply chains. Malhotra et al. 2018 proposed online search trends or e-commerce transactions as indicators to assess the speed of adoption of energy-efficient products.

5.13. SDG 13: Climate Action

SDG 13 agenda is to act to slow climate change and prepare for the bad things that will happen as a result. Climate changes include higher temperatures, less rain, and faster winds (Ziboon et al. 2022). Elias et al. 2020 utilized multi-source Earth observation methods for evaluating land deformation and sea level rise in Greek coastal regions, revealing noteworthy disparities in the alterations of relative sea levels, which are of utmost importance for developing climate change adaptation and mitigation strategies. Huntingford et al. 2019 utilized machine learning and artificial intelligence can aid in achieving a thorough comprehension and adjustment to the effects of climate change by revealing intricate relationships within the climate system and improving predictions of weather events.

5.14. SDG 14: Life Below Water

The fourteenth sustainable development goal is to preserve and utilize oceans, waters, and marine resources sustainably. For oil disaster remote sensing, infrared cameras analyze the thermal emissions of oil to identify it. On the other hand, laser fluor sensors and radar possess distinct qualities for accurately distinguishing oil and conducting extensive searches over wide areas, respectively (Fingas & Brown, 2014).

5.15. SDG 15: Life on Land

To protect, restore, and promote the sustainable use of terrestrial environments as part of SDG 15: dealing with trees in a way that does not harm them; combat desertification; stop and reverse land pollution; stop the loss of biodiversity. Sudhakar Reddy et al. 2016 utilized grid cell analysis on a multi-source and multi-temporal dataset to measure and track deforestation in India from 1930 to 2013 across a span of eight decades. The satellite images were obtained from the cloud-free Landsat Multispectral Scanner System (MSS) between 1972 and 1977, IRS 1A/IB LISS I in 1995, IRS P6 Advanced Wide Field Sensor (AWiFS) in 2005, and Resourcesat-2 AWiFS in 2013. The forest cover accuracy exceeded 89%. The application of GIS analysis on degraded landscapes in Adwa district, Ethiopia, utilizing LANDSAT ETM data, has identified connections between land-cover types, landform features, and main soil groups. This research has yielded crucial foundational data for effective land management in comparable places (Gala et al. 2011).

5.16. SDG 16: Peace, Justice, and Strong Institutions

SDG 16 aims to create peaceful and welcoming societies, ensure that everyone has equitable access to justice, and create institutions at all levels that are effective, accountable, and welcoming. The court systems are utilizing big data and associated technologies, albeit their usage is currently limited, to combat crime Završnik, 2021.

5.17. SDG 17: Partnerships for the Goals

SDG 17 aspires to build global collaborations to help achieve the SDGs. Naz et al. 2019 utilized new tools for cloud computing and handling large amounts of data to facilitate the formation of international collaborations focused on achieving the same SDGs and harnessing the potential of big data for sustainable development.

Table 1. Case studies related to SDGs used geospatial methods

Author(s)	Year	Methodology	Technique	Area of Interest	Findings
Singh et al.	2022	Multiple regression analysis	Statistical analysis	Saudi Arabia	Positive correlation between work, gender equality, education, and GDP. Negative correlation with poverty, food, and health metrics
Sami	2024	Geographic data analysis	Geospatial data tools	Sudan states	Geospatial data aids in accelerating investment processes
Hu et al.	2023	Remote sensing, geospatial data analysis	Remote sensing, statistical bulletin	Fujian Province, China	GEP increased by 27.9%. Organizational services contributed most to GEP, and GDP grew faster than GEP.
Choi et al.	2016	Open geospatial tools	Geospatial technology, open data platform	South Korea	Open technologies offer the potential for improving SDG measurements
Carneiro & Quinh-oneiro,	2015	Artificial intelligence (AI)	Fuzzy logic, AI models	Brazil	The AI model provides a holistic sustainability index, adaptable to other countries.
Shen,	2014	Urban Sustainability Indicator (USI) model	Satellite imagery, spatiotemporal analysis	Saskatoon, Saskatchewan, Canada	Insights for addressing urban geographical challenges
Watmough et al.	2016	Remote sensing, GIS analysis	Remotely sensed data, geographic metrics	Assam, India	Metrics such as land cover and travel time to market towns significantly influence welfare.
Sedda et al.	2015	Spatial and aspatial analysis	NDVI correlation, satellite data	West Africa	The inverse relationship between poverty and NDVI indicates the utility of satellite-based poverty modeling.
Kelly-Fair et al.	2022	Remote sensing, GIS analysis	Remote sensing, GIS	Semarang, Indonesia	Shift from forests to urban and plantation landscapes, guided by the SDGs for policy recommendations.
Nagy et al.	2018	GIS mapping, normalized quantitative data	GIS mapping, data normalization	Metrop-olitan Area, Romania	Progress varies across the region, with the municipality leading in SDG achievements.
Huan et al.	2019	Statistical analysis, geospatial analysis	GIS, statistical data	Central Asia (Kazak-hstan, Kyrgyzstan)	Highlights the need for improved data collection and research for sustainable development challenges

6. Case Studies of Geospatial Techniques in Sustainable Development

These case studies show how GIS and remote sensing are used in many industries to aid in sustainable development and decision-making. They emphasized the importance of these strategies in addressing global challenges. Sustainable development makes use of the geographic data and methodologies shown in Table 1.

7. Key Findings

The findings of this study clearly demonstrate the importance of geospatial research in understanding and addressing issues related to sustainable development. A key finding is the ability of geospatial analysis to clarify and understand how development challenges are spread across different regions. By analyzing spatial data, decision-makers can identify areas most vulnerable to environmental, social, and economic challenges. This insight promotes more efficient targeting of resources, ensuring that interventions are focused where the need is greatest. For example, maps that show poverty rates or track deforestation can help local governments prioritize their efforts.

Moreover, this spatial understanding is of great importance in monitoring progress towards achieving sustainable development goals. Geospatial tools provide the ability to visualize changes in indicators over time, giving a clear picture of whether specific goals, such as reducing poverty or promoting sustainable agriculture, are being achieved. This ongoing monitoring helps in adjusting strategies and assessing the impact of policies and programs. Geospatial research not only identifies development challenges but also plays a vital role in empowering communities.

Local groups can use spatial data to better understand their environments and actively participate in shaping the future of their regions. This empowerment comes through the provision of accessible and transparent information, which fosters rich public debate, encourages community engagement, and contributes to improved local governance. For example, communities that receive spatial data about their exposure to climate change or natural disasters are better able to advocate for preventive measures or develop contingency plans. Data-driven decision-making enhances the adaptive capacity of these communities, making them stronger and better prepared to face challenges. One of the key benefits of geospatial research is its ability to facilitate fact-based decision-making. Policymakers and development agencies can make more informed decisions using accurate, real-time data. For example, geospatial analytics in agriculture can provide insights into soil health, crop conditions, and weather patterns, helping farmers adopt sustainable practices and increase productivity while minimizing environmental damage.

Furthermore, geospatial tools foster cross-sectoral collaboration. Sustainable development goals often require input and collaboration from multiple fields, such as ecology, economics, and urban planning. Geospatial research provides a common platform for these sectors to exchange data, share findings, and work toward a common goal. For example, urban planners and environmental scientists can work together to design greener, more resilient cities by sharing spatial data on infrastructure, land use, and natural resources.

The findings of this study also highlight the importance of investing in geospatial infrastructure and skills. To make the most of the potential of geospatial research, governments and local institutions need to provide the right technological infrastructure, from satellite imagery to geographic information systems. In many cases, developing countries may lack the tools and expertise to effectively collect and analyze spatial data. Investing in strengthening local geospatial research capacity is, therefore, critical to achieving sustainable development goals. This includes developing training programs for local experts, providing access to modern technologies, and fostering an environment that supports the flourishing of geospatial research. Through these steps, countries can fully harness the power of spatial data, ensuring that development efforts are data-driven, effective, and sustainable.

Finally, geospatial research is a vital component in promoting collective action towards achieving the SDGs. By providing a clear picture of where interventions are needed and how effective they are, spatial data encourages all stakeholders, including governments, NGOs, and international donors, to coordinate their efforts. In addition, the availability of high-quality spatial data can attract more funding for development projects, as donors and investors are more likely to support initiatives backed by reliable data that demonstrates tangible impact. In short, geospatial research provides valuable tools for understanding and addressing complex sustainable development challenges. It enables more targeted interventions, strengthens community capacities, encourages cross-sectoral collaboration, and supports investment in infrastructure and skills. Harnessing these tools is essential for achieving the

SDGs and promoting inclusive, long-term, sustainable development around the world.

Conclusion

To understand and deal with complex development problems, geospatial analysis is a comprehensive tool. According to the declaration, actions must be targeted and specific to each location. Progress must be tracked and analyzed, and rewards must be evaluated using geospatial data. According to the study, geographic analysis is important for many reasons, such as making decisions based on facts, investing in data infrastructure and capacity development, giving local people more power, working together across sectors, and making things more flexible. Reaching sustainable growth goals will make the future better and more stable. Future research may look at creative uses of geospatial analysis in a variety of fields, including environmental management, urban planning, and disaster response.

References

Acharya, T. D., & Lee, D. H. (2019). Remote Sensing and Geospatial Technologies for Sustainable Development: A Review of Applications. *Sensors & Materials*, 31(11), 3931–3945. https://doi.org/10.18494/SAM.2019.2706.

Arora, N. K., & Mishra, I. (2019). United Nations Sustainable Development Goals 2030 and environmental sustainability: race against time. *Environmental Sustainability*, 2, 339-342. https://doi.org/10.1007/s42398-019-00092-y.

Asi, Y. M., & Williams, C. (2018). The role of digital health in making progress toward Sustainable Development Goal (SDG) 3 in conflict-affected populations. *International journal of medical informatics*, 114, 114-120. https://doi.org/10.1016/j.ijmedinf.2017.11.003.

Avtar, R., Aggarwal, R., Kharrazi, A., Kumar, P., & Kurniawan, T. A. (2020). Utilizing geospatial information to implement SDGs and monitor their progress. *Environmental monitoring and assessment,* 192, 35. https://doi.org/10.1007/s10661-019-7996-9.

Bachmann, N., Tripathi, S., Brunner, M., & Jodlbauer, H. (2022). The contribution of data-driven technologies in achieving the sustainable development goals. *Sustainability*, 14(5), 2497. https://doi.org/10.3390/su14052497.

Bonafoni, S., Baldinelli, G., & Verducci, P. (2017). Sustainable strategies for smart cities: Analysis of the town development effect on surface urban heat island through remote sensing methodologies. *Sustainable Cities and Society*, 29, 211-218. https://doi.org/10.1016/j.scs.2016.11.005.

Brussel, M., Zuidgeest, M., Pfeffer, K., & Van Maarseveen, M. (2019). Access or accessibility? A critique of the urban transport SDG indicator. *ISPRS International Journal of Geo-Information*, 8(2), 67. https://doi.org/10.3390/ijgi8020067.

Carneiro, Á., & Quinhoneiro, F. (2015). A methodology approach for analysis of sustainability indicators as a tool for decision making using fuzzy logic. In *Sustainable Development and Planning VII* (pp. 345-354), *WIT Transactions on Ecology and The Environment,* 193, https://doi.org/10.2495/SDP150291.

Choi, J., Hwang, M., Kim, G., Seong, J., & Ahn, J. (2016). Supporting the measurement of the United Nations' sustainable development goal 11 through the use of national urban information systems and open geospatial technologies: a case study of south Korea. Open Geospatial Data, *Software and Standards*, 1, 4. https://doi.org/10.1186/s40965-016-0005-0.

Costa, D. G., Bittencourt, J. C. N., Oliveira, F., Peixoto, J. P. J., & Jesus, T. C. (2024). Achieving Sustainable Smart Cities through Geospatial Data-Driven Approaches. *Sustainability*, 16(2), 640. https://doi.org/10.3390/su16020640.

Das, S., Ganguly, K., Mitran, T., & Chakraborty, S. D. (2023). Applications of geospatial and information technologies toward achieving sustainable development goals. In *Application of Remote Sensing and GIS in Natural Resources and Built Infrastructure Management* (pp. 1-27). Springer, Cham. https://doi.org/10.1007/978-3-031-14096-9_1.

Dell'Acqua, F., Iannelli, G. C., Torres, M. A., & Martina, M. L. (2018). A novel strategy for very-large-scale cash-crop mapping in the context of weather-related risk assessment, combining global satellite multispectral datasets, environmental constraints, and in situ acquisition of geospatial data. *Sensors*, 18(2), 591. https://doi.org/10.3390/s18020591.

Dhumad, S., Jasim, O., & Hamed, N. (2023). The evaluation of carbon monoxide emissions using regression analysis and GIS techniques. *IOP Conference Series: Earth and Environmental Science,* 1129, 012041. https://doi.org/10.1088/1755-1315/1129/1/012041.

Eguaroje, E. O., Omali, T. U., & Umoru, K. (2020). Assessment of Drought Occurrence in Kano State, Nigeria. *International Journal of Trend in Scientific Research and Development*, 4(2), 297-303.

Elias, P., Benekos, G., Perrou, T., & Parcharidis, I. (2020). Spatio-temporal assessment of land deformation as a factor contributing to relative sea level rise in coastal urban and natural protected areas using multi-source earth observation data. *Remote Sensing*, 12(14), 2296. https://doi.org/10.3390/rs12142296.

Elvidge, C. D., Sutton, P. C., Ghosh, T., Tuttle, B. T., Baugh, K. E., Bhaduri, B., & Bright, E. (2009). A global poverty map derived from satellite data. *Computers & Geosciences*, 35(8), 1652-1660. https://doi.org/10.1016/j.cageo.2009.01.009.

Fingas, M., & Brown, C. (2014). Review of oil spill remote sensing. *Marine Pollution Bulletin*, 83(1), 9-23. https://doi.org/10.1016/j.marpolbul.2014.03.059.

Gain, A. K., Giupponi, C., & Wada, Y. (2016). Measuring global water security towards sustainable development goals. *Environmental Research Letters*, 11(12), 124015. https://doi.org/10.1088/1748-9326/11/12/124015.

Gala, T. S., Pazner, M., & Beyene, S. (2011). Evaluating biophysical attributes of environmentally degraded landscapes in Northern Ethiopia using LANDSAT ETM data and GIS. *Ethiopian Journal of Environmental Studies and Management*, 4(1), 1-16. https://doi.org/10.4314/ejesm.v4i1.1.

Hashim, A. H., Jasim, O. Z., & Salih, M. M. (2021). Developing and establishing a geospatial database of the religious endowments in Hilla, Iraq. *IOP Conference Series: Earth and Environmental Science*, 754, 012026. https://doi.org/10.1088/1755-1315/754/1/012026.

Hashim, A. H., Jasim, O. Z., & Salih, M. M. (2022). The establishing of geospatial database for agricultural lands of Islamic WAQF in Iraq: Case study Babil Province. *IOP Conference Series: Earth and Environmental Science*, 961, 012025. https://doi.org/10.1088/1755-1315/961/1/012025.

Hassani, H., Huang, X., MacFeely, S., & Entezarian, M. R. (2021). Big data and the united nations sustainable development goals (UN SDGs) at a glance. *Big Data and Cognitive Computing,* 5(3), 28. https://doi.org/10.3390/bdcc5030028.

Hassani, H., Yeganegi, M. R., Beneki, C., Unger, S., & Moradghaffari, M. (2019). Big data and energy poverty alleviation. *Big Data and Cognitive Computing*, 3(4), 50. https://doi.org/10.3390/bdcc3040050.

Hilal, A., Jasim, O., & Ismael, H. (2021). Determination of the optimum number and distribution of the ground control points in stereo imaging to achieve precise positions. *Journal of Physics: Conference Series*, 1973(1), 012191. https://doi.org/10.1088/1742-6596/1973/1/012191.

Hilal, A., Jasim, O., & Ismael, H. (2022). Accuracy Assessment for points coordinates surveyed using low-cost Unmanned Aerial Vehicle and Global Positioning System with 3Dsurvey and 3DF Zephyr software. *IOP Conference Series: Earth and Environmental Science*, 961, 012046. https://doi.org.10.1088/1755-1315/961/1/012046.

Hu, Q., Lu, C., Chen, T., Chen, W., Yuan, H., Zhou, M., Qiu, Z., & Bao, L. (2023). Evaluation and analysis of the gross ecosystem product towards the sustainable development goals: A case study of Fujian Province, China. *Sustainability*, 15(5), 3925. https://doi.org/10.3390/su15053925.

Huan, Y., Li, H., & Liang, T. (2019). A new method for the quantitative assessment of Sustainable Development Goals (SDGs) and a case study on Central Asia. *Sustainability*, 11(13), 3504. https://doi.org/10.3390/su11133504.

Huntingford, C., Jeffers, E. S., Bonsall, M. B., Christensen, H. M., Lees, T., & Yang, H. (2019). Machine learning and artificial intelligence to aid climate change research and preparedness. *Environmental Research Letters*, 14(12), 124007. https://doi.org/10.1088/1748-9326/ab4e55.

Husein, H., Jasim, O., & Mahmood, S. (2018). Proposal of building a standard geodatabase for urban land use. *MATEC Web of Conferences*, 162, 03024. https://doi.org/10.1063/5.0140830.

Husein, H. N., Hamed, N. H., & Jasim, O. Z. (2023). A python-based model for the selection of suitable locations for installing large-scale photovoltaic system in western Iraq by using GIS-machine learning techniques. *AIP Conference Proceedings*, 2775(1), 050009. https://doi.org/10.1051/matecconf/201816203024.

Ilie, C. M., Brovelli, M. A., & Coetzee, S. (2019). Monitoring SDG 9 with global open data and open software–A case study from rural Tanzania. *The International Archives of the Photogrammetry, Remote Sensing and Spatial Information Sciences*, 42, 1551-1558. https://doi.org/10.5194/isprs-archives-XLII-2-W13-1551-2019.

Jasim, O., Al-Anbari, R., & Mohammed, Z. (2020). MODIS and Landsat satellite images comparison for modeling near-surface air temperature. *IOP Conference Series: Materials Science and Engineering*, 737, 012211. https://doi.org/10.1088/1757-899X/737/1/012211.

Jasim, O., Hamed, N., & Abdulgabar, T. (2018). Change detection and building spatial geodatabase for Iraqi marshes. *MATEC Web of Conferences*, 162, 03021. https://doi.org/10.1051/matecconf/201816203021.

Jasim, O. Z., Hassoon, K. I., & Sadiqe, N. (2019). Mapping LCLU Using Python Scripting. *Engineering and Technology Journal*, 37, 140-147. https://doi.org/10.30684/etj.37.4A.5.

Jean, N., Burke, M., Xie, M., Davis, W. M., Lobell, D. B., & Ermon, S. (2016). Combining satellite imagery and machine learning to predict poverty. *Science*, 353(6301), 790-794. https://doi.org/10.1126/science.aaf7894.

Karim, H. H., Ziboon, A. R. T., & Al-Hemidawi, L. M. (2016). Assessment of water quality indices for Shatt Al-Basrah River in Basrah city, Iraq. *Engineering and Technology Journal,* 34, 1804-1822. https://doi.org/10.30684/etj.34.9A.8.

Kelly-Fair, M., Gopal, S., Koch, M., Pancasakti Kusumaningrum, H., Helmi, M., Khairunnisa, D., & Kaufman, L. (2022). Analysis of land use and land cover changes through the lens of SDGs in Semarang, Indonesia. *Sustainability*, 14(13), 7592. https://doi.org/10.3390/su14137592.

Khalaf, A. Z., Alwan, I., & Kadhum, T. (2018). Accuracy Assessment of World View-2 Satellite Imagery for Planimetric Maps Production. *Engineering and Technology Journal,* 36(1), 1-9. https://doi.org/10.30684/etj.36.1A.1.

Kuffer, M., Wang, J., Nagenborg, M., Pfeffer, K., Kohli, D., Sliuzas, R., & Persello, C. (2018). The scope of earth-observation to improve the consistency of the SDG slum indicator. *ISPRS International Journal of Geo-Information,* 7(11), 428. https://doi.org/10.3390/ijgi7110428.

Laituri, M., Davis, D., Sternlieb, F., & Galvin, K. (2021). SDG Indicator 11.3. 1 and secondary cities: an analysis and assessment. *ISPRS International Journal of Geo-Information*, 10(11), 713. https://doi.org/10.3390/ijgi10110713.

Machiwal, D., Jha, M. K., & Mal, B. C. (2011). Assessment of groundwater potential in a semi-arid region of India using remote sensing, GIS and MCDM techniques. *Water resources management*, 25, 1359-1386. https://doi.org/10.1007/s11269-010-9749-y.

Malhotra, C., Anand, R., & Singh, S. (2018). Applying big data analytics in governance to achieve sustainable development goals (SDGs) in India. In *Data science landscape* (pp. 273-291). Springer, Singapore. https://doi.org/10.1007/978-981-10-7515-5_19.

Nagy, J. A., Benedek, J., & Ivan, K. (2018). Measuring sustainable development goals at a local level: A case of a metropolitan area in Romania. *Sustainability*, 10(11), 3962. https://doi.org/10.3390/su10113962.

Nashait, A. F., Jasim, O., Ismail, M., & Saad, F. (2020). Integrating various satellite images for identification of the water bodies through using machine learning: A case study of

Salah Adin, Iraq. *IOP Conference Series: Materials Science and Engineering*, 737, 012223. https://doi.org/10.1088/1757-899X/737/1/012223.

Naz, M., Al-zahrani, F. A., Khalid, R., Javaid, N., Qamar, A. M., Afzal, M. K., & Shafiq, M. (2019). A secure data sharing platform using blockchain and interplanetary file system. *Sustainability*, 11(24), 7054. https://doi.org/10.3390/su11247054.

Nicolau, R., David, J., Caetano, M., & Pereira, J. M. (2018). Ratio of land consumption rate to population growth rate—Analysis of different formulations applied to mainland Portugal. *ISPRS International Journal of Geo-Information*, 8(1), 10. https://doi.org/10.3390/ijgi8010010.

Oluboyede, T., & Odeyemi, C. (2020). The impact of geospatial information on sustainable development goals (SDGs) agenda 2063 in Africa. *IOSR Journal of Humanities And Social Science (IOSR-JHSS)* 25(8), 47-56. https://doi.org/10.9790/0837-2508154756.

Orimoloye, I., Mazinyo, S., Nel, W., & Kalumba, A. (2018). Spatiotemporal monitoring of land surface temperature and estimated radiation using remote sensing: human health implications for East London, South Africa. *Environmental earth sciences*, 77, 77. https://doi.org/10.1007/s12665-018-7252-6.

Palomares, I., Martínez-Cámara, E., Montes, R., García-Moral, P., Chiachio, M., Chiachio, J., Alonso, S., Melero, F. J., Molina, D., & Fernández, B. (2021). A panoramic view and SWOT analysis of artificial intelligence for achieving the sustainable development goals by 2030: progress and prospects. *Applied Intelligence*, 51, 6497-6527. https://doi.org/10.1007/s10489-021-02264-y.

Pandey, P. C., & Pandey, M. (2023). Highlighting the role of agriculture and geospatial technology in food security and sustainable development goals. *Sustainable Development*, 31(5), 3175-3195. https://doi.org/10.1002/sd.2600.

Saad, F., Jasim, O., & Albayati, M. (2023). Literature review of mobile spatial data collection and related to reality as a basis for the geomatics applications. *IOP Conference Series: Earth and Environmental Science*, 1129, 012009. https://doi.org/10.1088/1755-1315/1129/1/012009.

Sadiqe, N. E., Jasim, O. Z., & Al-Bakri, M. (2023). Geospatial Techniques for Preparing the Requirements of 3D Modeling for Smart City Planning-Review paper. *The Iraqi Journal of Architecture and Planning*, 22(2), 73-85. https://doi.org/10.36041/iqjap.2023.141103.1078.

Sami, K. A. (2024). Activating Geospatial Information for Sudan's Sustainable Investment Map. *International Journal of Trend in Scientific Research and Development*, 8(1), 904-916.

Sedda, L., Tatem, A. J., Morley, D. W., Atkinson, P. M., Wardrop, N. A., Pezzulo, C., Sorichetta, A., Kuleszo, J., & Rogers, D. J. (2015). Poverty, health and satellite-derived vegetation indices: their inter-spatial relationship in West Africa. *International health*, 7(2), 99-106. https://doi.org/10.1093/inthealth/ihv005.

Shelestov, A., Lavreniuk, M., Kussul, N., Novikov, A., & Skakun, S. (2017). Exploring Google Earth Engine platform for big data processing: Classification of multi-temporal satellite imagery for crop mapping. *Frontiers in Earth Science*, 5, 232994. https://doi.org/10.3389/feart.2017.00017.

Shen, L. (2014). *Monitoring urban sustainability based on an integrated indicator model using geospatial technique and multiple data sources: a case study in the city of Saskatoon*, Saskatchewan, Canada University of Saskatchewan.

Singh, H. P., Singh, A., Alam, F., & Agrawal, V. (2022). Impact of sustainable development goals on economic growth in Saudi Arabia: Role of education and training. *Sustainability*, 14(21), 14119. https://doi.org/10.3390/su142114119.

Sudhakar Reddy, C., Jha, C., Dadhwal, V., Hari Krishna, P., Vazeed Pasha, S., Satish, K., Dutta, K., Saranya, K., Rakesh, F., & Rajashekar, G. (2016). Quantification and monitoring of deforestation in India over eight decades (1930–2013). *Biodiversity and Conservation*, 25, 93-116. https://doi.org/10.1007/s10531-015-1033-2.

Trinidad, J. E. (2022). Spatial analysis of high school dropout: The role of race, poverty, and outliers in New York City. *SocArXiv Papers*. https://doi.org/10.31235/osf.io/9nwst.

Truby, J. (2020). Governing artificial intelligence to benefit the UN sustainable development goals. *Sustainable Development*, 28(4), 946-959. https://doi.org/10.1002/sd.2048.

Watmough, G. R., Atkinson, P. M., Saikia, A., & Hutton, C. W. (2016). Understanding the evidence base for poverty–environment relationships using remotely sensed satellite data: an example from Assam, India. *World Development*, 78, 188-203. https://doi.org/10.1016/j.worlddev.2015.10.031.

Wetter, E., Sundsøy, P. R., Engø-Monsen, K., Lu, X., Tatem, A. J., Bengtsson, L., Steele, J. E., Pezzulo, C., Alegana, V., & Bird, T. J. (2017). Mapping poverty using mobile phone and satellite data. *Interface*, 4(127). https://doi.org/10.1098/rsif.2016.0690.

Završnik, A. (2021). Algorithmic justice: Algorithms and big data in criminal justice settings. *European Journal of Criminology*, 18(5), 623-642. https://doi.org/10.1177/1477370819876762.

Ziboon, A. T., Albayati, M. M., & Dalhel, F. T. (2022). Monitoring soil degradation in the Mesopotamian Plain using GIS and remote sensing techniques. *Engineering and Technology Journal*, 40(5), 649-660. http://doi.org/10.30684/etj.v40i5.2121.

Chapter 7

An AI-Driven Solution for Addressing Climate Change Challenges

Wolfram Hardt
Shadi Saleh*
Batbayar Battseren
Reda Harradi
and Ariane Heller
Department of Computer Engineering, Faculty of Computer Science,
Chemnitz University of Technology, Chemnitz, Germany

Abstract

This chapter investigates the integration of Artificial Intelligence (AI) in enhancing climate resilience and promoting sustainable environmental solutions. It provides a comprehensive analysis of the current opportunities and challenges in this field, emphasizing the critical role of AI in addressing complex environmental issues. The research identifies significant bottlenecks in the practical application of AI technologies, highlighting the need for robust support services and specialized training programs to facilitate effective implementation. A specific use case demonstrates the application of AI in automated monitoring, illustrating its capabilities in processing large datasets and enabling real-time decision-making. The findings underscore the importance of centralized training initiatives to foster the widespread adoption of AI methodologies across various sectors. This research contributes to a deeper

* Corresponding Author's Email: shadi.saleh@informatik.tu-chemnitz.de

In: Advancements in Sustainability Systems II
Editors: H. H. Al-Kayiem, R. A. Jessam, Hasanain A. A. Wahhab et al.
ISBN: 979-8-89530-562-1

understanding of AI's potential in mitigating the impacts of climate change and advancing sustainable environmental practices.

Keywords: data monitoring, AI services, sustainable development

Abbreviations

AI	Artificial Intelligence
ANN	Artificial Neural Network
DL	Deep Learning
DNN	Deep Neural Network
ML	Machine Learning
UAV	Unmanned Aerial Vehicle
UAS	Unmanned Aerial System
MVS	Multi-Value Sensor
IoU	Intersection over Union
mAP	Mean Average Precision
DEM	Digital Elevation Model
IAISCC	International AI Services & Competence Centre
GPU	Graphics Processing Unit
QPB	Query Pool-Based Batch
GPS	Global Positioning System
YOLO	You Only Look Once

1. Introduction

The pressing challenge of climate change and the pursuit of climate resilience stand as critical issues within our contemporary civilization. The multifaceted nature of these challenges necessitates a comprehensive approach, integrating various domains into a cohesive system (Hariram et al. 2023, Ma et al. 2023). Technology plays a pivotal role in supporting this integration, enabling solutions that address the complexities of climate impacts while fostering sustainable development (Abbass et al. 2022, Raihan, 2023).

Biodiversity encompasses the variety of life forms on Earth, including terrestrial, freshwater, and marine species. It is a fundamental component of ecosystem health, providing essential services that sustain human life and well-being. Ecosystems must maintain suitable living conditions for all organisms to thrive, with factors such as habitat quality, resource availability, and climate stability being paramount. The decline in biodiversity, exacerbated by climate change, habitat destruction, and pollution, threatens these vital systems. Research indicates that biodiversity loss can destabilize ecosystems, leading to reduced resilience against environmental stressors (Islam et al. 2022, Sandifer et al. 2015).

Energy is indispensable to modern society, underpinning essential services such as home heating, cooking, lighting, and automation. The transition to renewable energy sources is crucial for mitigating climate change and ensuring sustainable energy access. Recent studies highlight the importance of transitioning from fossil fuels to renewables, emphasizing the potential of solar, wind, and hydroelectric power (Mutezo and Mulopo, 2021, and Hassan et al. 2024). Furthermore, energy efficiency measures and smart grid technologies are vital for optimizing energy consumption and reducing greenhouse gas emissions.

Efficient transportation systems are integral to societal functioning, facilitating the movement of people and goods across various distances. The ability to provide affordable and rapid transportation is essential for enhancing connectivity, supporting economic growth, and improving access to healthcare and education. In urban areas and developing countries, integrated transportation solutions can significantly contribute to sustainable development and smart city initiatives (Mishra et al. 2022). The adoption of electric vehicles and public transport systems powered by renewable energy sources is increasingly recognized to reduce the carbon footprint of transportation.

The quality of water and air is critical to human and animal health, influencing overall well-being and quality of life. Access to clean water and breathable air is fundamental for preventing diseases and promoting health (Leal Filho et al. 2020). Poor environmental quality can lead to significant public health challenges, including respiratory diseases and waterborne illnesses. Research suggests that improvements in water and air quality can lead to substantial health benefits, emphasizing the need for policies that prioritize environmental health (Adanma & Ogunbiyi, 2024). Ensuring high standards of water and air quality is essential for fostering peaceful coexistence among diverse populations globally.

Given the interconnected nature of these domains, significant investments in climate resilience and a sustainable energy future are imperative. Technological innovation is at the forefront of this endeavor. Besides optimizing existing production and consumption processes, there is a critical need for novel methods and approaches that leverage advanced technologies, particularly those driven by Artificial Intelligence (AI).

2. AI Addressing Climate Challenges

Artificial Intelligence (AI) has emerged as a transformative technological paradigm across diverse sectors, offering innovative solutions to complex climate-related challenges. The past decade has witnessed significant advancements in AI technologies, opening new avenues for research and practical applications, particularly in the following areas:

- Resource Management: AI algorithms, particularly those employing deep learning and reinforcement learning techniques, have demonstrated significant potential in enhancing the efficiency of resource utilization. This is particularly evident in energy systems, water management, and waste management through the implementation of predictive analytics and real-time monitoring systems (Matheri et al. 2022). These AI-driven approaches facilitate more precise resource allocation and consumption forecasting, leading to optimized utilization and reduced environmental impact.
- Biodiversity Conservation: Machine learning techniques, including convolutional neural networks and random forests, have proven effective in analyzing vast ecological datasets. These methods enable the identification of critical habitats, tracking of species populations, and informing evidence-based conservation strategies, thereby contributing to the mitigation of biodiversity loss (Hoffmann, 2022). The integration of AI with remote sensing and geographic information systems has further enhanced the spatial and temporal resolution of biodiversity monitoring efforts.
- Urban Planning and Smart Cities: AI-driven tools, such as agent-based models and genetic algorithms, facilitate data-driven urban planning processes. These technologies optimize transportation networks, reduce congestion, and promote sustainable practices through the analysis of complex urban systems (Ashwini et al. 2022). The implementation of AI in smart city initiatives has led to more

efficient resource allocation and improved quality of life for urban residents while minimizing environmental impact.

- Climate Adaptation: AI can assist in developing sophisticated predictive models that assess climate impacts and support communities in adapting to changing environmental conditions (Leal Filho et al. 2022). These models, often based on ensemble methods and Bayesian networks, integrate multiple data sources to provide robust projections of climate-related risks and inform adaptive strategies at various scales.

The integration of AI technologies in addressing climate challenges represents a significant advancement in environmental science and policy. However, further research is necessary to address challenges related to data quality, model interpretability, and the ethical implications of AI-driven decision-making in environmental contexts.

2.1. Key Challenges in Climate Resilience

Climate change presents an unprecedented global challenge, impacting diverse ecosystems and human societies across multiple scales. This multi-faceted issue necessitates robust technological interventions to mitigate its impacts and ensure sustainable trajectories for future generations. The integration of Artificial Intelligence (AI) within climate resilience and sustainable energy sectors holds significant potential to revolutionize environmental stewardship and resource management paradigms.

Despite the theoretical promise of AI applications in these domains, the practical assimilation of AI technologies has been characterized by sluggish adoption rates and fragmented implementation strategies. The complex interconnections between biodiversity, energy systems, transportation networks, and urban development necessitate a synchronized approach to effectively leverage AI capabilities. These systems are intrinsically dependent on air and water quality, which serve as critical determinants of both human health outcomes and biodiversity preservation.

The core problem lies in the suboptimal utilization of AI capabilities to integrate these diverse domains into a cohesive system that enhances predictive modeling, adaptive capacity, and responsive mechanisms to environmental challenges. This integration deficit manifests in several key areas:

- Insufficient data integration: The lack of standardized protocols for data collection, storage, and sharing across different environmental sectors impedes the development of comprehensive AI models.
- Limited cross-domain application: AI applications often remain siloed within specific environmental domains, failing to capture the complex interdependencies between different ecological and anthropogenic systems.
- Inadequate scaling of AI solutions: Many AI-driven environmental solutions demonstrate efficacy at local or regional scales but face significant challenges in scaling to address global climate issues.
- Suboptimal resource allocation: The underutilization of AI in resource management leads to inefficiencies in energy consumption, water usage, and waste management, exacerbating environmental stressors.
- Delayed adaptive responses: The absence of integrated AI systems hinders the development of real-time, adaptive strategies to address rapidly changing environmental conditions.

This multifaceted gap in AI integration and utilization significantly impedes the optimization of resource consumption, hampers the implementation of sustainable practices, and exacerbates the vulnerabilities of already stressed ecosystems. Addressing this problem requires a concerted effort to develop holistic, AI-driven approaches that can effectively synthesize diverse environmental data streams, generate actionable insights, and facilitate adaptive management strategies across multiple spatial and temporal scales.

2.2. Key Focus Areas of Research

The primary objective of this research is to catalyze the adoption of trustworthy Artificial Intelligence (AI) in enhancing climate resilience and promoting sustainable energy solutions. This undertaking involves leveraging the latest advancements in AI technologies, such as deep learning, machine learning, and neural networks, to develop methodologies that are not only theoretically robust but also practically applicable and scalable in real-world contexts.

This research seeks to innovate the integration of AI technologies with environmental systems, extending beyond mere application to ensure that ethical considerations, transparency, and accountability are embedded in AI deployment. Specifically, it aims to address the identified gap in AI applications, wherein the rapid advancements in AI capabilities and research outputs are not effectively translated into practical deployment, particularly within environmental applications.

By concentrating on the development of an integrated multi-value sensor concept and enhancing data quality through the utilization of balanced, clean, and representative datasets, this research will provide actionable insights and robust solutions that are directly applicable to pressing environmental challenges.

Moreover, through the implementation of detailed case studies and active learning strategies, this research will demonstrate the efficacy of AI-driven methodologies in transforming environmental monitoring and sustainable management practices. This objective is critical for fostering a high-quality, sustainable, and harmonious lifestyle while simultaneously advancing human and ecological health and longevity through improved environmental conditions.

3. AI Application Gap

The development of Artificial Intelligence (AI) methods has accelerated significantly in recent years. Traditional approaches primarily relieve artificial neural networks (ANNs), which are structured around an input layer, one or more hidden layers, and an output layer. Each layer processes information through weighted parameters, with a limited number of computing neurons and interconnections, resulting in relatively low computational complexity. These methods excel at recognizing straightforward patterns and performing interpolations.

In contrast, the next generation of AI methods introduced machine learning (ML) as a conceptual framework, marking a significant departure from traditional rule-based programming paradigms. ML algorithms empower systems to learn from data, identify patterns, and make predictions or decisions without the need for explicit programming instructions. This paradigm shift has revolutionized numerous fields, including image recognition, natural language processing, and autonomous vehicles.

Over the past decade, deep learning (DL) has emerged as a prominent AI approach, further enhancing the capabilities of machine learning. DL algorithms, particularly deep neural networks (DNNs), utilize multiple layers of interconnected neurons to extract hierarchical representations from raw data. This architecture enables DL models to learn intricate features and complex patterns, resulting in remarkable breakthroughs in areas such as computer vision, speech recognition, and drug discovery. Compared to traditional machine learning methods, deep learning techniques demonstrate superior performance in handling large volumes of unstructured data and achieving state-of-the-art results across diverse applications.

The proliferation of affordable high-performance computing platforms facilitates this rapid development. Computing centers are no longer the sole providers of computational power; developers and application engineers can now leverage GPU-based boards, such as those in the NVIDIA Jetson family (Valladares et al. 2021). These boards offer a compact and powerful platform specifically designed for embedded systems and AI applications. The accessibility of such platforms has made algorithm development and case studies feasible for a wide range of applications (Guo et al. 2022). This trend is evident in the increasing volume of AI research publications, projected to exceed 1,000 publications annually, as shown in Figure 1.

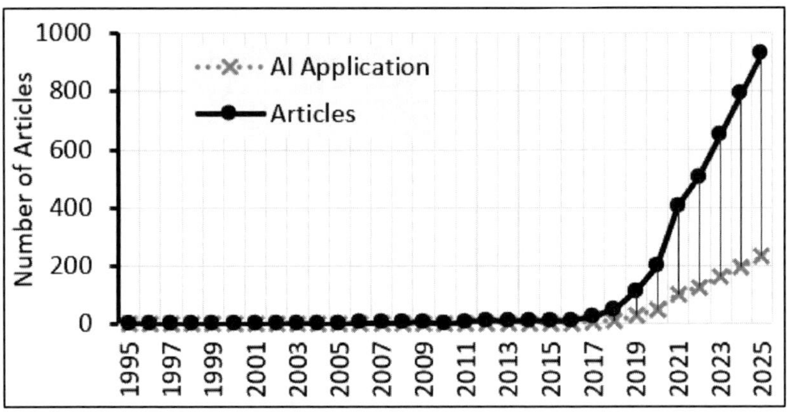

Source: Authors.

Figure 1. Ratio between published articles and deployed AI applications.

The surge in AI research activity has also prompted optimization engineers to apply these methods across various domains. Technological advancements have mitigated additional bottlenecks, particularly concerning memory requirements for storing sensed data for processing and analysis. As illustrated in Figure 2, the costs associated with sensors and memory have decreased, while the necessary computational power has become readily available (Khan & Zakarya, 2021). Consequently, the integration of data, memory, and computation enables the application of AI methods to address real-world environmental issues.

Despite these advancements, our investigations reveal that the number of successful AI-based real-world applications is not increasing to an anticipated extent. The disparity between the rapid growth of AI research outputs and the deployment of AI methods in practical applications is evident, which we term the AI-Application Gap.

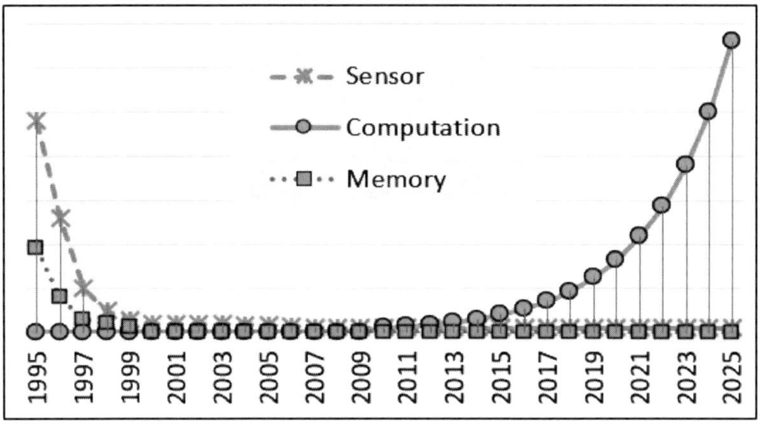

Source: Authors.

Figure 2. The declining cost of sensors and cloud memory versus the increase in computation complexity.

The AI application gap is critical for the effective deployment of the latest AI advancements in real-world scenarios, particularly in environmental applications, climate issues, and energy sustainability. Closing this gap requires substantial effort and resources. To this end, we are conducting further studies to identify and understand the new bottlenecks hindering the application of AI-based methods, especially as they pertain to environmental challenges.

4. Integrated Multi-Value Sensor Concept

To achieve environmental intelligence, the implementation of AI methods is essential, particularly due to the substantial amount of sensor data generated.

Our research emphasizes that data is a critical component, necessitating consideration of the following aspects.

4.1. Data Collection

The sensing of environmental issues requires a huge number of sensors. Typical applications, such as air quality monitoring, water quality monitoring, or monitoring of geo-technical stability, e.g., for early detection of hang sliding, have to cover large areas with a size of several square kilometers. Thus, hundreds or thousands of sensors are needed. Optimized wireless sensor networks can support data collection (Ho et al. 2005). In other cases, mobile sensors can be used. If sensors are mounted on cars or unmanned aerial vehicles, the sensors will become mobile. This reduces the number of sensors but also increases the complexity of data collection at the same time, due to the added mobility.

4.2. Data Quality

High data quality is crucial for developing AI applications as it directly impacts the performance, reliability, and effectiveness of the models. Some studies, e.g., Cichy & Rass, 2020, offer an overview of data quality frameworks, comparing methodologies for defining, assessing, and improving data quality across diverse business environments. They emphasize the challenges in selecting suitable frameworks and provide a decision guide to assist organizations in choosing appropriate methods to maintain high data quality standards.

Quality data ensures that AI algorithms can learn accurate patterns and make informed decisions, leading to more reliable predictions and outcomes. Without high data quality, AI models may produce erroneous results, introduce biases, or fail to generalize correct output with respect to new data. Additionally, clean and relevant data reduces the risk of overfitting and improves the interpretability of AI models, making them more transparent and trustworthy. Ultimately, investing in high data quality is essential for maximizing the potential of AI applications and achieving meaningful results in various domains. High data quality is paramount for developing successful and reliable AI applications. The most essential criteria for achieving high data quality in AI development include:

- Balance: Balance refers to the even distribution of data across different classes, categories, or labels within the dataset. Imbalanced data, where one class is significantly more prevalent than others, can lead to biased AI models that perform poorly on minority classes. Ensuring balance in the dataset helps maintain model fairness and accuracy across all classes.
- Cleanliness: Clean data is free from errors, inconsistencies, and irrelevant information. Noise, outliers, missing values, and duplicate entries can adversely affect the performance of AI models by introducing inaccuracies and biases. Data cleaning processes such as pre-processing, outlier detection, and deduplication are essential for ensuring data cleanliness.
- Representativeness: Representativeness ensures that the dataset accurately reflects the diversity and variability of the real-world phenomena it aims to capture. A representative dataset contains samples that adequately represent the underlying population, demographics, or conditions. Without representativeness, AI models may fail to generalize well with respect to unseen data, leading to poor performance in real-world scenarios.

By prioritizing criteria such as balance, cleanliness, and representativeness, developers can ensure that their AI applications are built on high-quality data, leading to more accurate, reliable, and effective outcomes.

For real-world scenarios, many heterogenous sensors have to be connected. Relevant data for specialized monitoring is needed, along with huge amounts over time, in order to compute situation awareness, system robustness, health monitoring, and all kinds of estimations of a system's future behaviors and statuses. In our concept, we integrated several sensors, sensing different kinds of data, especially temperature, light, and velocity, as well as imaging and 2D/3D LiDAR data. These sensors can be considered as one integrated multi-value-sensor (MVS). In addition, such MVS can be mounted on mobile platforms such as cars for the exploration of cities and urban regions. MVS can also be mounted on unmanned aerial vehicles (UAVs) in order to leverage air mobility, e.g., in inspection tasks. In our use case, we are using MVS on micro air vehicles, a subclass of UAVs.

In all cases, MVS data is collected in a cloud-based system. Here, we use a modular web-based approach with open interfaces and data visualization methods. Details can be found in (Bergelt & Englisch, 2020). On the cloud server side, AI-based methodologies are deployed. Initially, automated AI algorithms assess the quality of data, with emphasis placed on factors such as data volume and balance. Subsequently, the extraction of training data undergoes meticulous processing. Our services adhere to stringent selection criteria to facilitate an optimized training phase for various types of AI algorithms. Finally, the deployment of trained algorithms to the real-world application comes into view, as illustrated in Figure 3.

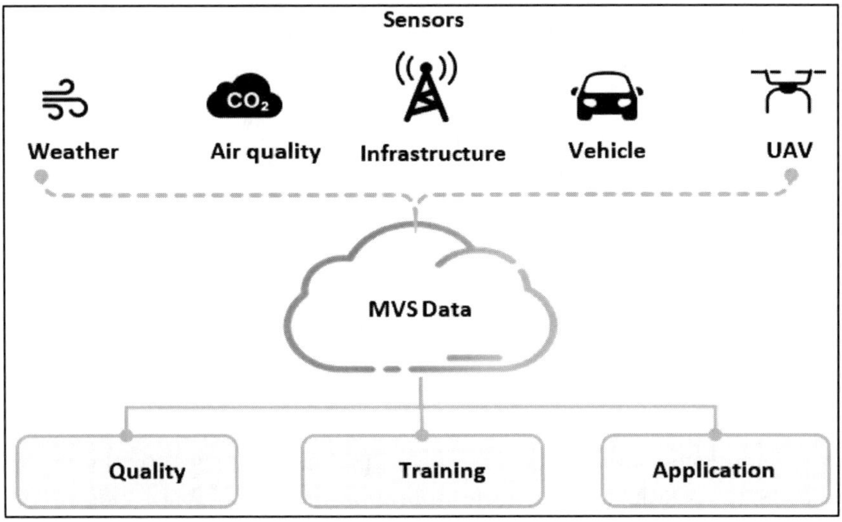

Source: Authors.

Figure 3. Real-world MSV scenario with data-oriented AI services.

The concept depicted can be applied to any application generically. This makes it adaptable and very powerful. Our concept has been evaluated in two use cases.

5. Forest Inspection Case Study

In recent years, insect infestation has significantly contributed to deforestation in Germany. The Federal Statistical Office of Germany reports a concerning rise in wood damage caused by insects, with the bark beetle being a significant culprit. In recent data, bark beetles have been linked to the destruction of approximately 60.1 million cubic meters of timber. Notably, in 2020 alone, this damage represented a staggering 72% of all wood affected by pests throughout Germany (Destatis, 2021). This has prompted the authorities to employ certain methods to mitigate its impact. However, these conventional methods rely on labor-intensive forest patrolling routines to identify infected trees, a process that is time-consuming and prone to leaving areas uninspected due to the size and density of the forest. Conversely, UAV-based inspections offer a solution capable of covering large areas in a short timeframe, leveraging aerial imagery to detect infected trees. For instance, Lehmann et al. (2015) present an approach using unmanned aerial system (UAS) and color infrared images to monitor pest infestation levels in forestry, specifically targeting the oak splendor beetle and its impact on oak trees.

This chapter presents a remote sensing-based forest inspection solution utilizing a deep learning approach (Battseren et al. 2022). The result demonstrates the potential of a deep learning-assisted method to enable efficient large-scale UAV-based forest monitoring, as in Figure 4. The proposed approach consists of three main phases: Data collection, AI model training, and Geo-localization.

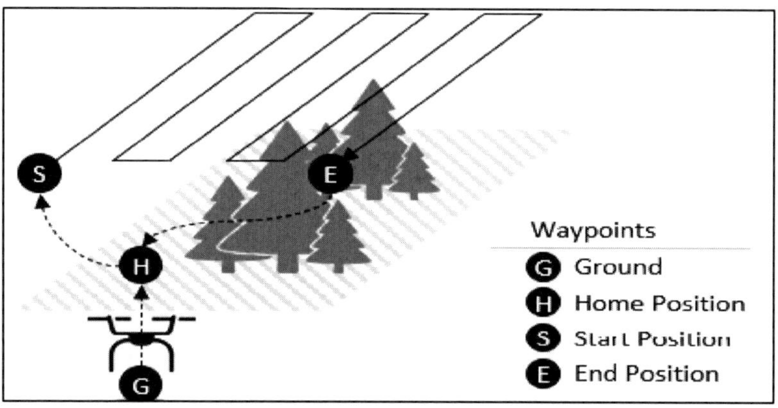

Source: Authors.

Figure 4. UAV and AI-based forest inspection.

5.1. Data Collection

Data collection is performed using a Yuneec E90 camera mounted on a UAV, capturing inspection videos in 4K pixel resolution at 30 frames per second. Initial videos were recorded in different locations and at different times of the day to diversify the training data, and experts analyzed them. These videos have been used for preparing image datasets for AI model training. Expert feedback was essential for identifying and labeling the infested trees.

In real-world inspection, the same video data must be collected to localize the infested trees. In addition to visual information, these videos include geolocation data of the flight path, updated every 200 milliseconds.

5.2. AI Model Training

A deep learning-based approach is employed for tree detection, with an active learning method (Saleh et al. 2019) utilized to mitigate data scarcity during the model training.

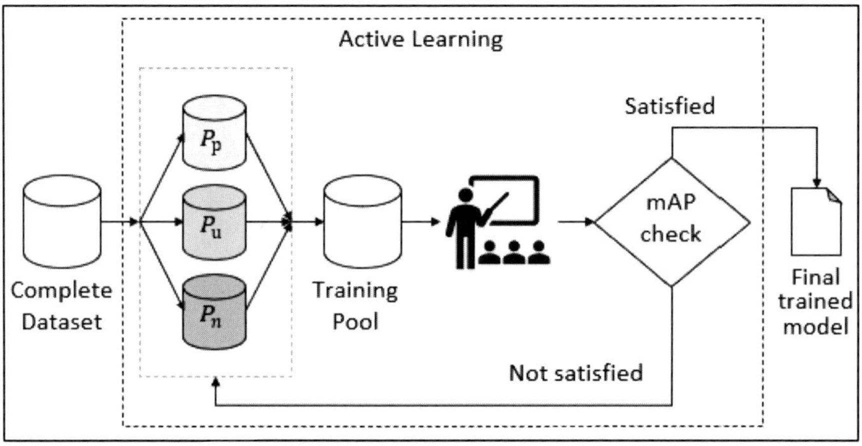

Source: Authors.

Figure 5. Query pool-based batch learning method.

YOLOv4 was selected for detecting the infested trees. Based on the initial dataset, a total of 1,552 labeled images were prepared and divided into training (70%), validation (10%), and test (20%) sets. A query pool-based batch (QPBB) learning method was employed as an active learning approach to achieve comparable accuracies to the conventional learning strategies while utilizing a minimal set of training data. While the conventional model used the entire training dataset, the active learning strategy initially employed only 15% of the images, increasing by 4% with each iteration. Additionally, the training dataset was categorized based on identifiability and was divided into three pools: "Positive" (35%), with clearly identifiable objects; "Unsure" (45%), with identifiable but less clear objects; and "Negative" (20%), with hardly identifiable objects. This approach aims to enhance training efficiency by focusing on the most informative data, as depicted in Figure 5.

To assess the efficiency of the proposed model, it was evaluated against a conventional model based on the mean Average Precision (mAP) metric and the classification loss score. The mAP was calculated considering an Intersection over a Union (IoU) threshold of 0.5. Table 1 presents the results of training the model using a Tesla T4 GPU. After 17 iterations, utilizing 56% of the training dataset, i.e., 599 images, the model achieved a mAP score of 63.5, compared to 62.7 from the conventional model.

Table 1. Comparison of training the model using active learning

Iteration	Used Dataset	Used Dataset [%]	Precision	Recall	mAP @0.5	mAP @0.5-0.95
1	161	15%	0.317	0.324	0.201	0.063
2	197	18%	0.140	0.603	0.332	0.107
3	232	22%	0.120	0.730	0.363	0.112
4	266	25%	0.179	0.688	0.421	0.146
5	298	28%	0.232	0.622	0.421	0.138
6	329	31%	0.370	0.547	0.444	0.148
7	358	33%	0.340	0.536	0.430	0.164
8	387	36%	0.226	0.752	0.515	0.197
9	415	39%	0.251	0.679	0.448	0.147
10	441	41%	0.367	0.613	0.490	0.179
11	466	43%	0.353	0.659	0.537	0.202
12	491	46%	0.381	0.654	0.561	0.206
13	515	48%	0.298	0.730	0.553	0.197
14	537	50%	0.412	0.676	0.574	0.220
15	559	52%	0.468	0.623	0.555	0.199
16	579	54%	0.400	0.686	0.565	0.189
17	599	56%	0.293	0.790	0.635	0.231
18	619	58%	0.369	0.692	0.566	0.198
19	637	59%	0.428	0.639	0.571	0.213
20	1,076	100%	0.440	0.737	0.627	0.244

5.3. Geo-Localization

Geo-localization involves two primary processes: tree detection and localization. The tree detection model processes the input video and produces an output file with the detection results. In the previous section, the object detection part is explained, and the trained model is used for detecting the infested trees.

The localization process maps the detected trees from the video frames in the Cartesian coordinate system to real-world geographic coordinates. This task is performed frame by frame and is complicated due to the tree projection and the camera's perspective, as shown in Figure 6. A common triangulation technique can be used to calculate the tree's location. However, it requires the UAV's position, including the height above ground level. The problem is that the height information is not available due to the uneven ground level and tree height. As a solution, this information is derived from the Digital Elevation Model (DEM), which provides the ground surface height above sea level. However, it introduces inaccuracies, as it provides a single tree top elevation value for each 50 x 50 m area.

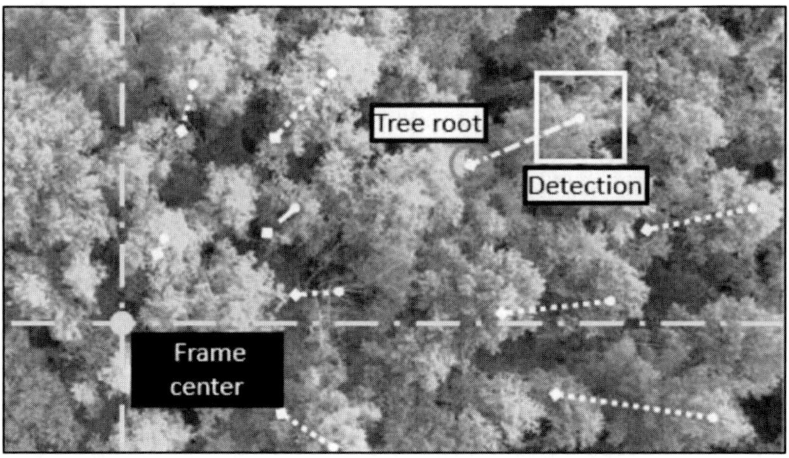

Source: Authors.

Figure 6. Height error source of the tree detection.

The test results show that the accuracy of the coordinates depends on the height of the trees and the UAV's altitude. Trees can appear with varying lengths extending to the side of the frame, depicted in dashed lines in Figure 5, causing the localized coordinates to often represent the ground obscured by the tree canopy, represented with circular symbols, rather than the tree root, i.e., the diamond symbols in Figure 6. These factors contribute to inaccuracies in the tree localization process.

A single tree appears in several image frames, each localized with a certain error rate due to the DEM. Hence, to obtain the final coordinates, the calculated coordinates were clustered using the DBSCAN algorithm. (Figure 7).

Finally, the proposed system demonstrates efficacy in detecting dying and deceased trees through aerial inspection, with the ability to geolocate them at an average error rate of 4.59 meters. The utilization of unmanned aerial vehicles for forest inspection offers significant advantages to forestry services compared to conventional foot patrolling methods.

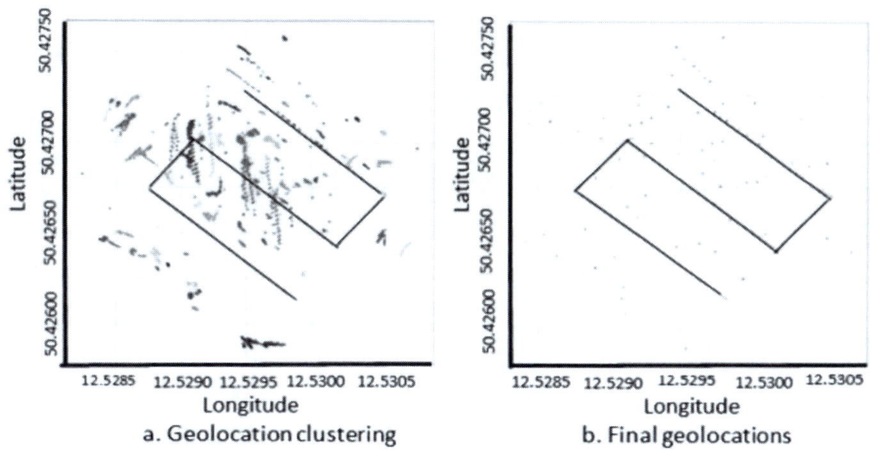

Source: Authors.

Figure 7. Tree localization result.

6. International AI Services Centre

Through the utilization of use cases, we have demonstrated the efficacy of AI methods in yielding high-quality results when data quality is upheld. Within our International AI Services & Competence Centre (IAISCC), we are dedicated to investigating fully automated services, such as web services for data evaluation. Furthermore, our AI Dashboard serves as a tool for visualizing data quality and assessing the suitability of AI algorithms for specific datasets. IAISCC operates as an open cooperation platform, facilitating the integration of contributions from various international partners.

Conclusion

Artificial Intelligence offers significant potential for addressing climate resilience and sustainable energy challenges. While AI research has advanced rapidly, the gap between the research and the practical application remains. Ensuring high-quality data is essential for successful AI implementation, with criteria such as balance, cleanliness, and representativeness being crucial. Our proposed integrated sensors concept facilitates environmental monitoring by combining various sensors into a unified system. Through case studies like forest inspection, we demonstrate the practical application of AI methods. At the International AI Services & Competence Centre, we aim to bridge the gap between AI research and use-case applications by developing automated services and fostering collaboration among international partners.

References

Abbass, K., Qasim, M. Z., Song, H., Murshed, M., Mahmood, H., & Younis, I. (2022). A review of the global climate change impacts, adaptation, and sustainable mitigation measures. *Environmental Science and Pollution Research*, 29(28), 42539-42559. https://doi.org/10.1007/s11356-022-19718-6.

Adanma, U. M., & Ogunbiyi, E. O. (2024). The public health benefits of implementing environmental policies: A comprehensive review of recent studies. *International Journal of Applied Research in Social Sciences*, 6(5), 978-1004. https://doi.org/10.51594/ijarss.v6i5.1148.

Ashwini, B. P., Savithramma, R. M., & Sumathi, R. (2022, May). Artificial intelligence in smart city applications: An overview. In 2022 6th *International Conference on Intelligent Computing and Control Systems* (ICICCS) (pp. 986-993). IEEE. https://doi.org/10.1109/ICICCS53718.2022.9788152.

Battseren, B., Harras, M. S., Aguirre, D. A. O., Saleh, S., & Hardt, W. (2022). Tree detection and localization approach for UAV-based forest inspection. *Embedded Self-Organizing Systems, 9*(3), 73-81. https://doi.org/10.14464/ess.v9i3.557.

Bergelt, R., & Englisch, N. (2020). Towards cloud-supported automotive software development and test. *Embedded Self-Organising Systems*, 7(2), 8-12. https://doi.org/10.14464/ess.v7i2.440.

Cichy, C., & Rass, S. (2019). An overview of data quality frameworks. *IEEE Access*, 7, 24634-24648. https://doi.org/10.1109/ACCESS.2019.2899751.

Destatis. (2021, August). *Forest damage: Logging of timber damaged by insect infestation grew more than tenfold within five years*. Federal Statistical Office. https://www.destatis.de/EN/Press/2021/08/PE21_N050_41.html.

Guo, Q., Ren, M., Wu, S., Sun, Y., Wang, J., Wang, Q., & Chen, Y. (2022). Applications of artificial intelligence in the field of air pollution: A bibliometric analysis. *Frontiers in Public Health*, 10, 933665. https://doi.org/10.3389/fpubh.2022.933665.

Hassan, Q., Viktor, P., Al-Musawi, T. J., Ali, B. M., Algburi, S., Alzoubi, H. M., & Jaszczur, M. (2024). The role of renewable energy in global energy transformations. *Renewable Energy Focus*, 48, 100545. https://doi.org/10.1016/j.ref.2024.100545.

Hariram, N. P., Mekha, K. B., Suganthan, V., & Sudhakar, K. (2023). Sustainalism: An integrated socio-economic-environmental model to address sustainable development and sustainability. *Sustainability*, 15(13), 10682. https://doi.org/10.3390/su15131 0682.

Ho, C. K., Robinson, A., Miller, D. R., & Davis, M. J. (2005). Overview of sensors and needs for environmental monitoring. *Sensors*, 5(1), 4-37. https://doi.org/10.3390/ s5010004.

Hoffmann, S. (2022). Challenges and opportunities of area-based conservation in reaching biodiversity and sustainability goals. *Biodiversity and Conservation*, 31(2), 325-352. https://doi.org/10.1007/s10531-021-02340-2.

Islam, M. M., Chowdhury, M. A. M., Begum, R. A., & Amir, A. A. (2022). A bibliometric analysis on the research trends of climate change effects on economic vulnerability. *Environmental Science and Pollution Research*, 29(39), 59300-59315. https://doi.org/ 10.1007/s11356-022-20028-0.

Khan, A. A., & Zakarya, M. (2021). Energy, performance, and cost-efficient cloud datacentres: A survey. *Computer Science Review*, 40, 100390. https://doi.org/10.1016 /j.cosrev.2021.100390.

Leal Filho, W., Wall, T., Azul, A. M., Brandli, L., & Özuyar, P. G. (Eds.). (2020). *Good health and well-being.* Cham: Springer International Publishing. https://doi.org/10. 1007/978-3-319-95681-7_300068.

Leal Filho, W., Wall, T., Mucova, S. A. R., Nagy, G. J., Balogun, A. L., Luetz, J. M., & Gandhi, O. (2022). Deploying artificial intelligence for climate change adaptation. *Technological Forecasting and Social Change*, 180, 121662. https://doi.org/10. 1016/j.techfore.2022.121662.

Lehmann, J. R. K., Nieberding, F., Prinz, T., & Knoth, C. (2015). Analysis of unmanned aerial system-based CIR images in forestry-A new perspective to monitor pest infestation levels. *Forests, 6*(3), 594-612. https://doi.org/10.3390/f6030594.

Ma, R., Abid, N., Yang, S., et al. (2023). From crisis to resilience: Strengthening climate action in OECD countries through environmental policy and energy transition. *Environmental Science and Pollution Research*, 30, 115480–115495. https://doi.org/ 10.1007/s11356-023-29970-z.

Matheri, A. N., Mohamed, B., Ntuli, F., Nabadda, E., & Ngila, J. C. (2022). Sustainable circularity and intelligent data-driven operations and control of the wastewater treatment plant. *Physics and Chemistry of the Earth, Parts A/B/C*, 126, 103152. https://doi.org/10.1016/j.pce.2022.103152.

Mishra, R. K., Kumari, C. L., Janaki Krishna, P. S., & Dubey, A. (2022). *Smart cities for sustainable development: An overview*. Chapter in the book Smart Cities for Sustainable Development (pp. 1-12). https://doi.org/10.1007/978-981-16-7410-5_1.

Mutezo, G., & Mulopo, J. (2021). A review of Africa's transition from fossil fuels to renewable energy using circular economy principles. *Renewable and Sustainable Energy Reviews*, 137, 110609. https://doi.org/10.1016/j.rser.2020.110609.

Raihan, A. (2023). A review of the global climate change impacts, adaptation strategies, and mitigation options in the socio-economic and environmental sectors. *Journal of Environmental Science and Economics*, 2(3), 36-58. https://doi.org/10.56556/jescae.v2i3.587.

Saleh, S., Khwandah, S. A., Mumtaz, A., Heller, A., & Hardt, W. (2019). Traffic signs recognition and distance estimation using a monocular camera. In *Proceedings of the 6th International Conference on Actual Problems of System and Software Engineering* (pp. 407-418). https://ceur-ws.org/Vol-2514/paper104.pdf.

Sandifer, P. A., Sutton-Grier, A. E., & Ward, B. P. (2015). Exploring connections among nature, biodiversity, ecosystem services, and human health and well-being: Opportunities to enhance health and biodiversity conservation. *Ecosystem Services*, 12, 1-15. https://doi.org/10.1016/j.ecoser.2014.12.007.

Valladares, S., Toscano, M., Tufiño, R., Morillo, P., & Vallejo-Huanga, D. (2021). Performance evaluation of the Nvidia Jetson Nano through a real-time machine learning application. In Intelligent Human Systems Integration 2021: *Proceedings of the 4th International Conference on Intelligent Human Systems Integration* (IHSI 2021), (pp. 343-349). Springer International Publishing. https://doi.org/10.1007/978-3-030-68017-6_51.

Chapter 8

The Effect of Induced Vibrations Caused by Internal Two-Phase Flow in Piping Systems

Hussein N. Jabbar[1,*]
Kifah H. Hilal[1]
and Zena K. Kadhim[2]
[1]Middle Technical University, Technical Engineering College, Baghdad, Iraq
[2]Department of Mechanical Engineering, Wasit University, Wasit, Iraq

Abstract

The phenomenon of two-phase flow in pipes has received great attention due to its importance in various industrial processes such as nuclear power generation, petroleum transportation, and chemical engineering. When two different phases, such as a liquid gas or a solid gas, flow simultaneously inside a confined tube, they interact dynamically, resulting in complex fluid-structure interactions that cause breakage and fatigue in the piping system because of vibration resulting from the movement of these fluids. This chapter provides a comprehensive analysis of vibrations induced by two-phase flow in pipes, focusing on the basic mechanisms, experimental observations, and engineering implications. Theoretical models and computational simulations are used to illustrate the system dynamics, considering factors such as flow regime transitions, phase distribution, and piping geometry. Experimental investigations, including vibration measurements and visualization techniques, provide valuable insights into the behavior of vibrations induced by two-phase flow under different operating conditions.

---

* Corresponding Author's Email: abc0071@mtu.edu.iq

In: Advancements in Sustainability Systems II
Editors: H. H. Al-Kayiem, R. A. Jessam, Hasanain A. A. Wahhab et al.
ISBN: 979-8-89530-562-1

Furthermore, the engineering implications of these vibrations are discussed, including their effect on piping integrity, material fatigue, and system performance. Understanding vibrations generated by two-phase flow is essential to improve system design, enhance operational safety, and mitigate potential risks in industrial applications.

Keywords: flow-induced vibrations, two-phase flow, piping systems, fluid-structure interaction

Abbreviations

FIV	Flow-induced vibration
FSI	Fluid-structure interaction
VIV	Vortex-induced vibration
CFD	Computational fluid dynamics

1. Introduction

The investigation of vibrations induced by two-phase flow in pipes is a critical area of study due to its relevance in various engineering applications. Several research studies have delved into this phenomenon, aiming to understand the characteristics and effects of two-phase flow-induced vibration (FIV) on piping systems. Miwa et al. 2015 researched two-phase flow-induced vibration in piping systems, shedding light on the complexities and implications of such vibrations. Additionally, Asiegbu et al. 2022 explored the effects of pipe diameter on internal multiphase flow-induced vibration at pipe elbows, emphasizing the importance of understanding how different parameters influence vibration characteristics. Furthermore, Miwa et al. 2016 developed a model capable of predicting force fluctuations and dominant frequency ranges induced by stratified wavy two-phase flow, providing insights into predicting FIV with accuracy. Moreover, the study by Bamidele et al. 2019 on two-phase flow-induced vibration of piping structures with flow-restricting orifices highlighted the significance of considering different flow conditions inducing vibrations. Additionally, Ibrahim et al. 2010 addressed fluid-elastic instability in single- and two-phase flows, emphasizing the need to comprehend the instability behavior induced by internal flows. Adegoke et al. 2020 investigated the nonlinear vibrations of cantilevered pipes conveying pulsating two-phase flow, contributing to the understanding of instability behaviors in pipes due to internal two-phase flow.

Vibrations caused by two-phase flow in piping systems are a major concern in the engineering industry. Vibrations can hinder the smooth operation of engineering devices and can cause serious consequences, such as system failure. This phenomenon has been investigated in both industrial and academic fields, with special emphasis on advances in two-phase flow-induced vibration research. Vibrations are caused by parametric excitation and resonance. Vibrations are studied in different flow patterns, such as champagne, toroidal, and toroidal, and the acceleration frequency response contains useful information about the flow. It was also studied in nonlinear transverse vibrations of a cantilever tube conveying a two-phase pulsating flow. Vibrations are expected to occur due to the change in the velocity of the transported fluid over time.

Vibrations are studied using the multiscale analysis method, where they are studied in the context of multiphase flow, which refers to the concurrent movement of substances in two or more thermodynamic phases. The field of multiphase flow is closely connected to the advancement of fluid mechanics and thermodynamics. The study of two-phase flow has a historical background and has been classified into both industrial and academic fields (Yamagata et al. 1997; Hara et al. 1976; Miwa et al. 2015). Vibrations caused by two-phase flow in pipes have been the subject of many studies. Ortiz et al. 2013 provided an in-depth discussion on FIV phenomena due to two-phase pipe flow. Adeshina et al. 2020 developed an experimental stand to investigate the influence of a two-phase flow on pipeline vibration under different flow patterns of a coolant. They also describe the procedure for studying the properties of developed and manufactured swirl models. Ortiz et al. 2013 collected a set of 32 two-phase horizontal flow cases and found that the acceleration frequency response contains useful information about the flow. Ebrahimi et al. 2019 proposed a mathematical model for the dynamic analysis of two-phase flow in vertical pipes and concluded that pipes conveying two-phase flow are prone to experiencing many dynamic phenomena. Zhou et al. 2016 studied the vortex-induced vibration of curved flexible tubes in shear flows and found that different vibration frequencies exist at different positions of the tube while presenting a multi-mode response. Muhammad et al. 2020 experimental study of flow-induced vibration of an in-water pipe. It is concluded that the flow-induced vibration effect due to the pipe conveying fluid increases the maximum deflection when the fluid velocity increases.

2. The Applications of Multiphase Flow Systems

Issues related to FIV are not considered during the design phase. These problems only appear when the system is fully operational and practical. As mentioned previously, the external indicators of FIV are noise and system vibrations. The most serious impacts can be small or large-scale incidents (Liu et al. 2013). Two-phase flows and the consequences of them inside or over surfaces could be classification as shown in the diagram, Figure 1.

The flow regime has a clear impact on FIV. In bubbly flows, the Source of the two-phase FIV is the vibration generated by the bubbles. This is different from momentum-induced oscillations in that momentum changes are not important for champagne flows. Likewise, momentum fluctuations are found to exhibit significant size in the slug flow phase. Slug flow is a phenomenon that takes place in internal flows where two phases are present. It is characterized by a distinct separation between the liquid and gaseous phases, with liquid "slugs" alternating with "Taylor bubbles" in the gaseous phase of the flow. The slug flow regime exhibits distinct and prominent variations in density. Significant momentum fluctuations are observed at pipe bends, where the centrifugal force component amplifies the changes in force.

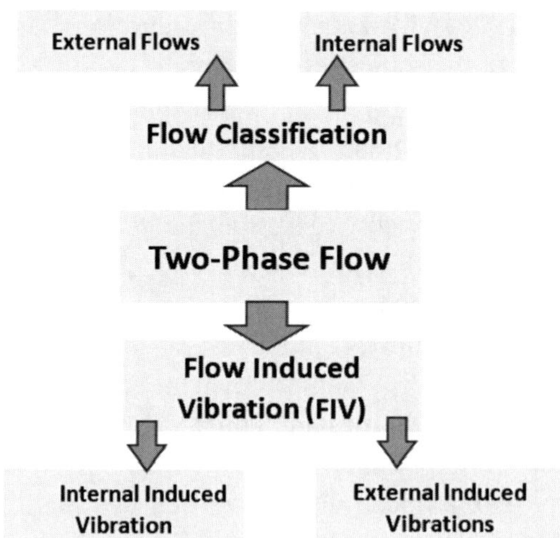

Source: Authors.

Figure 1. Classification of two-phase flows and flow-induced vibration by two-phase sources.

Table 1. Samples of studies of internal flow-induced vibrations

Author	Title	Findings
Dai et al. (2012)	Vibration analysis of three-dimensional pipes conveying fluid with consideration of steady combined force by the transfer matrix method.	The natural frequencies, frequency response functions, and instability of 3D-shaped pipeline systems are analyzed.
Mohmmed et al. (2019)	Effect of slug flow frequency on the mechanical stress behavior of pipelines	Data were analyzed statistically using SPSS software, and the concept of slug unit cell was adopted to deduce the developed model.
Al-Hashimy, et al. (2014)	Effect of various fluid densities on vibration characteristics in variable cross-section pipes	The motion for a pipe conveying fluid was derived using beam theory and mathematics.
Heng et al. (2022).	Investigation on two-phase flow-induced vibrations of a piping structure with an elbow.	Results show that phase separation and intense secondary flow occur at the elbow because of the centrifugal force, which results in the transition of the flow regime.
Mohmmed et al. (2020)	One-way coupled fluid-structure interaction of gas-liquid slug flow in a horizontal pipe: Experiments and simulations	At constant superficial liquid velocity, the maximum induced stresses on the pipe wall increased with increasing slug length and slug velocity. While for the slug frequency, the maximum principal stresses decreased with increasing slug frequency.

Vibrations caused by internal axial flow may be easily identified, as stated by Kandasamy et al. 2016. His results concluded that the slug length and pressure increase with increasing air superficial velocity, and the slug initiation position shifts closer to the inlet. The pressure gradient of the flow increases with increasing gas flow rate due to a higher mean velocity. Al-Hashimy et al. 2015 study provides insights into the influence of flow-induced vibration on local flow parameters in two-phase flow. Hibiki et al. 1998 experimental investigation of two-phase (air-water) flow characteristics in a horizontal steel pipe at atmospheric conditions. Todkar et al. 2016 identified different flow patterns or regimes and their corresponding void fractions in the pipe. The excitation force is strongly dependent on the momentum flux of two-phase flows and the impact of liquid slugs on the elbow (Liu et al. 2012). Investigation of excitation of the horizontal bending structure tube induced by two-phase flow. The effect of velocities on the pressure and lateral and axial vibration responses of the bending pipe can be calculated. The result showed that the pipe vibration amplitude increases with increasing mass flow, as well as the vertical vibration affected by the flow pattern. As the gas velocity increases, the amplitude of the vibration of the tube increases (Heng et al.

2022). A summary of the literature on FIV by internal flows is shown in Table 1.

3. Two-Phase Flow

The two-phase flow refers to the simultaneous flow of two immiscible fluids of gas and liquid. The flow of two immiscible liquids, such as oil and water, is commonly referred to as a two-component flow. Also, Two-phase flow refers to the concurrent movement of both gas and liquid phases for the same fluid. For most two-phase flow situations, it is crucial to differentiate between laminar and turbulent flow. The flow conditions of each phase may be determined by calculating the Reynolds number for each phase using superficial velocity, $U_{ks,}$ and the diameter, *D,* of the pipe (Wongwises et al. 1998)

$$Re_{ks} = \frac{\rho_k U_{ks} D}{\mu_k} \tag{1}$$

Laminar flow is assumed for superficial Reynolds number < 2000. The subscript *k* refers to liquid or gas.

3.1. Two-Phase Flow Patterns

The Baker flow pattern map for air-water flow is depicted in Figure 2. The axes are defined in terms of $(U_{Gs} \times \rho_G)/\lambda$, and $(U_{Ls} \times \rho_L/\lambda$, where $U_{Gs} \times \rho_G = G_g$ = mass flux of gas phase (gas mass flow rate/tube cross-sectional area) and $U_{Ls} \times \rho_L = G_g$ = mass flux of liquid phase (kg/m^2 s (liquid mass flow rate/tube cross-sectional area).

$$\lambda = \left(\frac{\rho_G \rho_L}{\rho_{air}\rho_{water}}\right)^{1/2} \tag{2}$$

$$\psi = \frac{\sigma_{water}}{\sigma}\left|\frac{\mu_L}{\mu_{water}}\left(\frac{\rho_{water}}{\rho_L}\right)\right|^{1/3} \tag{3}$$

where σ is surface tension (N/m), and σ_{water} = 0.072 N/m is the surface tension of air-water.

The Baker map works well for water/air and oil/gas mixture in tubes with smaller diameters, not bigger than 50 mm (Baker, 1954).

On the other hand, based on experimental measurements, Oshinowo and Charles, 1974 suggested that the flow patterns depend on the volumetric flow rate of the gas flow and other fluid dynamic properties. Their flow pattern map is shown in Figure 3.

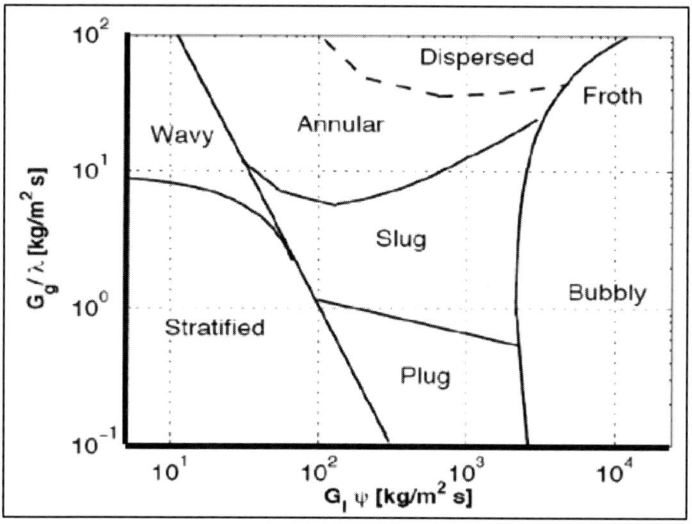

Figure 2. Baker flow pattern map for horizontal flow in a tube.

Oshinowo and Charles, 1974 have developed two maps, a. for upward flows in a pipe and b. for downward flows in a pipe. The flow pattern could be estimated through Eqs. (4) to (6).

$$R_v = \frac{Q_G}{Q_l} = \frac{U_G}{U_L} \tag{4}$$

$$Fr_{TP} = \frac{(U_G+U_L)^2}{g \times D} \tag{5}$$

$$A = \frac{\mu_L}{\mu_w}\left[\frac{\rho_L}{\rho_w}\left(\frac{\sigma_L}{\sigma_w}\right)^3\right]^{-0.25} \tag{6}$$

where Q_G is the volumetric flow rate of the gas, Q_L is the volumetric flow rate of the liquid, Fr_{TP} is the Froude number of the two phases, U_G and U_L are the gas and liquid superficial velocities, respectively.

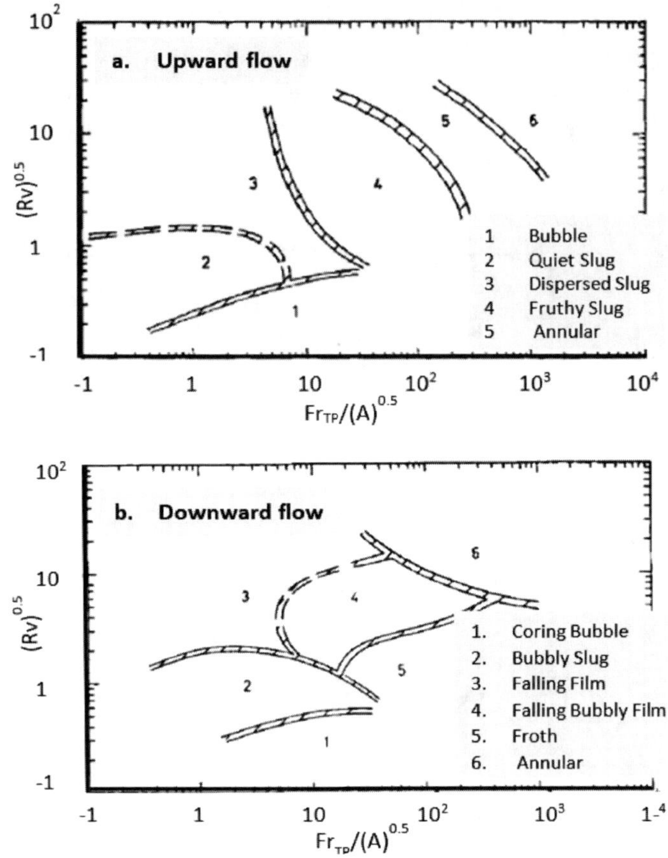

Source: Authors modified from Oshinowo and Charles, 1974.

Figure 3. Oshinowo and Charles' flow pattern map for flow in vertical pipes, a. Upward flow, b. Downward flow.

3.3. Description of Patterns of Flow Regimes

Brennen, 2005, and Crowe, 2006, provided an overview of multiphase flows in general, whereas Azzopardi, 2006 specifically focused on two-phase flows. Wörner, 2003 offers a comprehensive overview of numerical modeling for multiphase flows. Ghajar, 2005 gives a very informative tutorial specifically focused on two-phase flows that do not include phase transition. The flow patterns in horizontal two-phase flow differ from those in vertical two-phase flow due to the perpendicular action of gravity on the flow direction. The complexity of two-phase flow in horizontal tubes surpasses that of vertical two-phase flow due to the absence of axisymmetry caused by gravity. The flow patterns observed in horizontal two-phase flow are characterized as follows (Hewitt, 1998; Thome, 2004).

Table 2. Visualization of flow patterns in the vertical pipe flow regime

Type of flow	Description	Pattern outline and shape
Bubbly flow	When the gas flow rates are low, the gaseous phase is evenly distributed throughout the fluid, and the gas bubbles are consistently sized.	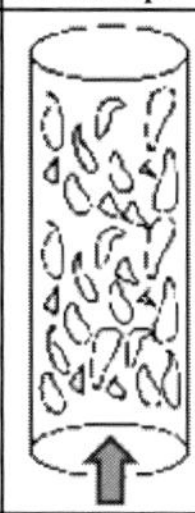
Slug flow	At greater gas flow rates, the collision and coalescence of bubbles lead to the formation of significantly larger bubbles known as "Taylor Bubbles." The liquid slugs exhibit a pattern of alternating with Taylor bubbles and typically contain bubbles in the gas phase.	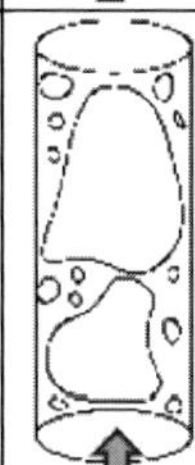
Churn flow	When the gas flow rate increases, the two-phase mixture is agitated, leading to the disintegration of the Taylor bubbles. The gas bubbles that have formed are now asymmetrical but substantial in size. The liquid phase also comprises minute gas bubbles.	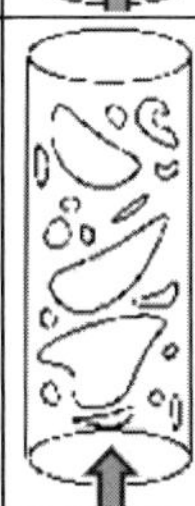
Annular flow	Characterized by the presence of a liquid layer on the walls of the pipe. The film is maintained by a central core of gas that is traveling at a high velocity. The gas phase often transports little liquid droplets.	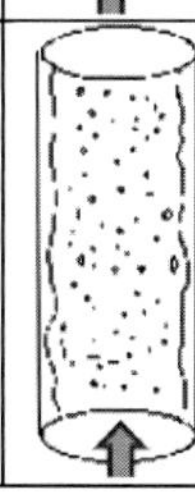

Source: Authors.

Table 3. Flow patterns in a horizontal pipe

Type of flow	Description	Pattern outline and shape
Stratified flow	Characterized by a distinct separation between the two phases, with the gas phase occupying the upper half of the pipe. Minor undulations are widespread along the boundary between the two stages.	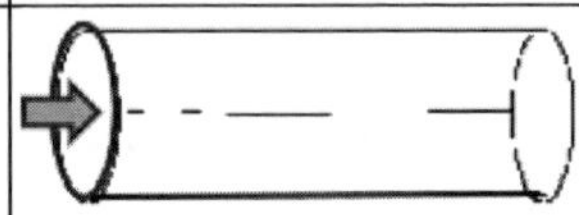
Slug flow	The horizontal flows have comparable characteristics to the slug flow found in vertical flows.	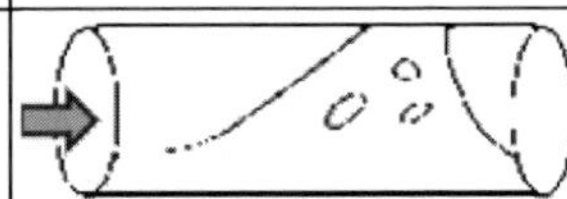
Annular flow	Analogous to the annular flow regime encountered in vertical, two-phase flows. Nevertheless, the liquid layer that sticks to the walls of the pipe is not evenly distributed, unlike in vertical flows.	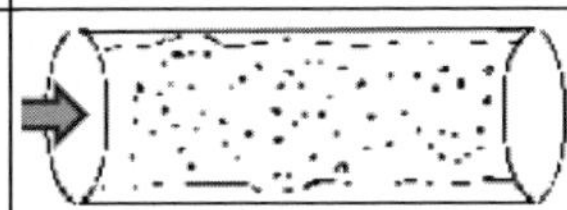
Bubbly flow	Exhibits similarities to vertical flows. Gas bubbles tend to ascend towards the upper section of the pipe.	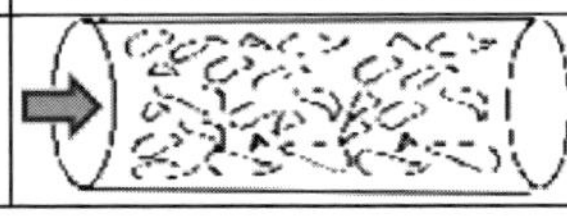

Source: Authors.

Vertically upward two-phase flows are important two-phase flow systems, which are well illustrated at a constant liquid flow rate, as shown in Table 2. They mainly exist in boiler flows, wall pipes in the power plant boilers, and the nuclear water reactors.

The horizontal two-phase flows are illustrated in Table 3. Gravity results in the gaseous phase appearing above the liquid phase.

4. Flow-Induced Vibrations

FIV refers to the phenomenon where structures experience oscillations due to the fluid flow around them. These vibrations can occur in various engineering systems, such as pipelines, aircraft wings, bridges, and offshore structures. FIV can lead to fatigue failure, reduced lifespan of the structure, and, in extreme cases, catastrophic failure (McNeil et al. 2016).

There are several mechanisms through which FIV can occur, including vortex shedding, galloping, flutter, and buffeting. Vortex shedding, for example, happens when alternating vortices form and shed from the structure, causing it to oscillate. Galloping occurs when the structure experiences a combination of lift and drag forces, resulting in large amplitude oscillations. Flutter involves the interaction between the structure's natural frequencies and the fluid flow, leading to self-excited oscillations. Predicting FIV requires a multidisciplinary approach involving fluid dynamics, structural mechanics, and vibration analysis. Researchers used CFD simulations, experimental testing, and theoretical modeling to study FIV and develop pacification strategies. Li et al. 2011Curried out an analysis of the free vibration for a piping system using the Timoshenko A model based on beam theory is proposed to analyze the dynamics of pipes carrying fluid. It is recommended to use the approach of dynamic stiffness to analyze the free vibration of multi-span pipes.

Several studies have investigated the vibration characteristics of pipelines that transport a mixture of gas and liquid. Hara et al. 1977 provided an analysis, both experimental and theoretical, of the process that causes vibrations in a straight horizontal tube when there is a flow of air and water in two phases. The reported works on FIV in various units of the piping system are summarized in Table 4.

Table 4. Summary of studies that dealt with various geometric shapes to study the effect of vibration-induced two-phase flow

Reference	Geometry model	Flow Direction	Section
Yang et al. (2023)	Spiral tube	Vertical	Numerical
Heng et al. (2022)	Elbow	Horizontal	Experimental
Mohmmed, et al. (2021)	Horizontal Pipes	Horizontal	Experimental
Reference	Geometry model	Flow Direction	Section
Mack et al. (2018)	90° bend	Horizontal	Numerical
Al-Kayiem, et al. (2017)	Horizontal Pipes	Horizontal	Experimental
Cargnelutti. et al. (2016)	T-Joint	Horizontal	Numerical
Reference	Geometry model	Flow Direction	Section
Al-Hashimy, et al. (2016)	Horizontal Pipes	Horizontal	Numerical
Giraudeau, et al. (2013)	U-Bend	Vertical	Numerical

4.1. Consequences of Vibrations Induced by Two-Phase Flow

The investigation of vibrations induced by two-phase flow in pipes is a critical area of study due to its relevance in various engineering applications. Several research studies have delved into this phenomenon, aiming to understand the characteristics and effects of two-phase FIV on piping systems. Miwa et al. 2022 discussed some of the effects of vibrations induced by two-phase flow:

4.1.1. Mechanical Stress and Fatigue

Vibrations caused by a two-phase flow can subject the pipe to mechanical stress and fatigue, potentially leading to structural damage or failure over time.

4.1.2. Flow Instabilities

Two-phase flow can cause vibrations to trigger flow instabilities such as slug flow, flow-induced vibrations, and flow regime transitions, affecting the overall system performance.

4.1.3. Heat Transfer Enhancement

Vibrations can enhance heat transfer rates in heat exchangers by promoting better mixing of the two phases and increasing the convective heat transfer coefficient.

4.1.4. Noise Generation

Vibrations induced by a two-phase flow can also generate noise, which may impact nearby equipment or personnel safety.

Table 5. Summary of the effect of two-phase flow vibration induced on the structure of horizontal and vertical pipes

References	Flow Direction	Research Highlights	Comments
Tay and Thorpe (2004)	Horizontal	The effect of liquid viscosity and liquid surface tension on the force acting	The reduction of liquid hold-up is associated with the decrease in liquid surface tension.
Liu et al. (2012)	Vertical	Negligible effects of friction	Input requirements: The parameters of interest are the pressure at the inlet and outlet.
Al-Hashimy, et al. (2016)	Horizontal	The volume of fluid method is employed, assuming unsteady, immiscible air-water flow.	The pressure gradient of the flow increases with increasing gas flow rate due to a higher mean velocity.
Miwa et al. (2014)	Vertical	Isothermal mixture	Application for a vertical elbow against gravity and horizontal.
Riverin, et al. (2007)	Vertical	Vibration response, excitation forces, and fluctuating properties of two-phase flow	The transducer was affixed to only one side for the course of the experiment.
Charreton, et al. (2015)	Vertical	Two-phase flow damping in pipes	The flow pattern influences the effectiveness of two-phase damping and is directly related to the void percentage in the bubbly flow regime.

Studies on FIV were mostly performed on piping systems with vertical or horizontal orientation, as summarized in Table 5.

4.2. Measurement Techniques of Vibrations Induced by Two-Phase Flows

Two-phase flow imaging is necessary to look at and understand the flow system. A CT scan. A well-established approach for describing flow that incorporates visualization of sections during use. Penetrating waves comprise X-rays, nuclear, optical, ultrasonic, and microwave. The technique of X-ray visualization has been widely employed for imaging two-phase flows. Aliseda et al. 2021 provide a comprehensive and commendable analysis of the X-ray imaging technique used for studying two-phase flows. X-ray tomography involves the passage of X-rays through a medium consisting of two phases, resulting in the attenuation of the X-rays. Beam attenuation is utilized to calculate the integral of local density along the course of the beam. Different beams are utilized at various angles to rebuild a picture that displays the density distribution of the phases. Table 6 displays details of the selected two-phase FIV.

Table 6. Summary of techniques for measuring vibrations induced by two-phase flow in pipes

Reference	Pipe direction, Two-phase	Pressure, Dynamic force measurement technique
Wang et al. (2002)	Horizontal	Pressure transducers are located at the intake and exit of the bend, and a quartz force sensor is also used.
Belfroid et al. (2010)	Horizontal	A dynamic pressure transducer is a type of force sensor.
Ortiz-Vidal et al. (2014)	Horizontal	Accelerometer
Al-Hashimy et al. (2015)	Horizontal	pressure drop prediction
Belfroid et al. (2016)	Horizontal-Bend-Vertical	Force Ring, Strain Gauge, Piezo Strain Gauge
Bamidele et al. (2021)	Horizontal, Flow Restricting Orifice	Capacitance Sensor, Laser Vibration Sensors

Conclusion

With increases in frequency and amplitude, the vibrational characteristic of frictional pressure drops becomes more intense. The pressure drops under the vibrational state are bigger than those under the non-vibrational state. Moreover, the pressure drop values relate to the fluid velocity, transverse vibration frequency, and transverse vibration amplitude. The variation characteristic of the air-water pressure drop is strongly affected by flow velocity. The pressure drop enlarges as the flow velocity becomes larger. The fluid flows in different flow patterns, such as slug, churn, and annular. Some of these flow patterns can generate dynamic fluid forces, which in turn may induce structural vibration. Excessive vibration may lead to component failures, which must be avoided as they can produce significant economic, environmental, and even human losses. Therefore, the knowledge of flow conditions, including flow patterns as well as flow-induced vibration (FIV), can have a significant impact on the proper design and operation of piping systems. First, the effect of pipe diameter upon FIV. Increasing the diameter of the pipe increases the excitation forces whilst decreasing the dominant frequency. However, the response of large-diameter pipes differs considerably from that of smaller-diameter pipes.

While working with vertical cylindrical two-phase flow, where gravity head is important, pipes were characterized by Schlegel et al. as large or small based on the non-dimensional hydraulic diameter. Afterward, the magnitude of the pipe's vibration reduces, but the frequency of the vibration increases. Computational Fluid Dynamics simulations enable the exploration of many subjects and facilitate the comparison of design alternatives that would otherwise be impractical or arduous to study through experimentation. From a conservative standpoint, it is challenging to determine if the real vibrations at the worst place are significantly greater than those at the measured locations. Simulations can somewhat mitigate this difficulty. A significant portion of the data needed to forecast pipe displacement is obtained via feedback and pertains to specific situations. Additionally, adhering to geometric design requirements will also mitigate the likelihood of vibration.

References

Adegoke, A., Fashanu, A., Adewumi, O., and Oyediran, A. (2020). Nonlinear vibrations of a cantilevered pipe conveying pulsating two phase flow. *Research on Engineering*

Structures and Materials, 6(3): 207-227. https://doi.org/10.17515/resm2019.145 me0902.

Al-Hashimy, Z. I., Al-Kayiem, H. H., Hasan, F., and Mohmmed, A. O. (2014). Effect of various fluid densities on vibration characteristics in variable cross-section pipes. *Journal of Applied Sciences*, 14: 2054-2060. http://dx.doi.org/10.3923/jas.2014. 2054.2060.

Al-Hashimy, Z. I., Al-Kayiem, H. H., Kadhim, Z. K., & Mohmmed, A. O. (2015). Numerical simulation and pressure drop prediction of slug flow in oil/gas pipelines. *WIT Trans. Eng. Sci.: Computational Methods Multiphase Flow VIII, 89*, 57-65. https://doi.org/10.2495/MPF150051.

Al-Hashimy, Z. I., Al-Kayiem, H. H., Nasif, M. S., and Mohmmed, A. O. (2016). Numerical simulation of the transient development of slug flow in horizontal pipes. *Applied Mechanics and Materials*, 819, 300-304. http://dx.doi.org/10.4028/www. scientific.net/AMM.819.300.

Al-Hashimy, Z. I., Al-Kayiem, H. H., Time, R. W., and Kadhim, Z. K. (2016). Numerical characterisation of slug flow in horizontal air/water pipe flow. *International Journal of Computational Methods and Experimental Measurements,* 4 (2), 114-130. http://dx.doi.org/10.2495/CMEM-V4-N2-114-130.

Aliseda, A., and Heindel, T. J. (2021). X-ray flow visualization in multiphase flows. *Annual Review of Fluid Mechanics,* 53(1), 543–567. https://doi.org/10.1146/annurev-fluid-010719-060201.

Al-Kayiem, H. H., Mohmmed, A. O., Al-Hashimy, Z. I., and Time, R. W. (2017). Statistical assessment of experimental observation on the slug body length and slug translational velocity in a horizontal pipe, *International Journal of Heat and Mass Transfer*, Vol. 105, 252–260. http://dx.doi.org/10.1016/j.ijheatmasstransfer.2016.09.105.

Asiegbu, N., Hossain, M., Droubi, G., & Islam, S. (2022). Investigation of the effects of pipe diameter of internal multiphase flow on pipe elbow vibration and resonance. *Proceedings of the Institution of Mechanical Engineers Part E, Journal of Process Mechanical Engineering*, 237(4), 1319-1330. https://doi.org/10.1177/09544089221 115520.

Azzopardi, B. J. (2006). *Gas-liquid flows*. Begell House- UK. https://doi.org/10.1615/978-1-56700-233-1.0.

Baker, O. (1954). Design of pipelines for the simultaneous flow of oil and gas. Paper presented at *the Fall Meeting of the Petroleum Branch of ASME*, Dallas, Texas, October 1953. https://doi.org/10.2118/323-G.

Bamidele, O. E., Hassan, M., and Ahmed, W. H. (2021). Flow induced vibration of two-phase flow passing through orifices under slug pattern conditions. *Journal of Fluids and Structures*, 101, 1. https://doi.org/10.1016/j.jfluidstructs.2020.103209.

Bamidele, O., Ahmed, W., and Hassan, M. (2019). Two-phase flow induced vibration of piping structure with flow restricting orifices. *International Journal of Multiphase Flow*, 113, 59-70. https://doi.org/10.1016/j.ijmultiphaseflow.2019.01.002.

Belfroid, S. P. C., Cargnelutti, M. F., Schiferli, W., van Osch M. (2010). Forces on bends and T-joints due to multiphase flow. In *Proceedings of the ASME 2010 3rd joint US-European fluids engineering Summer meeting Montreal*, Canada. August 1-5, 2010. 613–619. https://doi.org/10.1115/FEDSM-ICNMM2010-30756.

Belfroid, S. P. C., Nennie, E., and Lewis, M. (2016). Multiphase forces on bends – Large scale 6-inch experiments. In *Proceedings of the SPE annual technical conference and exhibition*, SPE-181604- M. S., Dubai, UAE. 26-28 September. https://doi.org/10.2118/181604-MS.

Brennen, C. E. (2005). *Fundamentals of multiphase flow*. Cambridge University Press. https://doi.org/10.1017/CBO9780511807169.

Cargnelutti, M. F., Belfroid, S. P. C., Schiferli, W., Van Osch, M. (2016). Multiphase Fluid Structure Interaction in Bends and T-Joints. In *Proceedings of the ASME 2010 Pressure Vessels and Piping Division/K-PVP Conference*, Bellevue, DC, USA, 18–22 July; pp. 1–8. https://doi.org/10.1115/PVP2010-25696.

Charreton, C., Béguin, C., Ross, A., Étienne, S., & Pettigrew, M. J. (2015). Two-phase damping for internal flow: Physical mechanism and effect of excitation parameters. *Journal of Fluids and Structures*, 56, 56–74. https://doi.org/10.1016/j.jfluidstructs.2015.03.022.

Crowe, C. T. (Ed.). (2006). *Multiphase flow handbook*. Taylor & Francis. https://doi.org/10.1201/9781420040470.

Dai, H. L., Wang, L., Qian, Q., & Gan, J. (2012). Vibration analysis of three-dimensional pipes conveying fluid with consideration of steady combined force by transfer matrix method. *Applied Mathematics and Computation*, 219(5), 2453–2464. https://doi.org/10.1016/j.amc.2012.08.081.

Ebrahimi-Mamaghani, A., Sotudeh-Gharebagh, Reza Zarghami, R., Mostoufi, N. (2019). Dynamics of two-phase flow in vertical pipes. *Journal of fluid and structures*, 2019. https://doi.org/10.1016/j.jfluidstructs.2019.03.010.

Ghajar, A. J. (2005). Non-boiling heat transfer in gas-liquid flow in pipes: A tutorial. *Journal of the Brazilian Society of Mechanical Sciences and Engineering*, 27 (1). https://doi.org/10.1590/S1678-58782005000100004.

Giraudeau, M., Mureithi, N. W., Pettigrew, M. J. (2013). Two-Phase Flow-Induced Forces on Piping in Vertical Upward Flow: Excitation Mechanisms and Correlation Models. *Journal of Pressure Vessels Technology, ASME Transaction*, 135, 1–16. https://doi.org/10.1115/1.4024210.

Hara, F. (1977). Two-phase-flow-induced vibrations in a horizontal piping system. *Bull. JSME* (Japan. Society of Mechanical Engineering) 20 (142), 419–427.

Hara, F. (1976). Two-Phase-Flow-Induced Vibrations in A Horizontal Pulping System. *Jsme International Journal Series B-fluids And Thermal Engineering*, 1976. https://doi.org/10.1299/jsme1958.20.419.

Heng, S. U., Yegao, Q. U., Wang, G., and Peng, Z. (2022). Investigation on two-phase flow-induced vibrations of a piping structure with an elbow. Applied Mathematical Mechanic, 43(11), 1657–1674. https://doi.org/10.1007/s10483-022-2916-6.

Hewitt, G. F. (1998). Multiphase Fluid Flow and Pressure Drop. In Heat Exchanger Design Handbook, Vol. 2, Begell House, New York, NY.

Hibiki, T., & Ishii, M. (1998). Effect of flow-induced vibration on local flow parameters of two-phase flow. *Nuclear Engineering and Design, 185*(2-3), 113-125. https://doi.org/10.1016/S0029-5493(98)00241-6.

Ibrahim, R. (2010). Overview of mechanics of pipes conveying fluids—part i: fundamental studies. *Journal of Pressure Vessel Technology*, 132(3). http://dx.doi.org/10.1115/1.4001271.

Kandasamy, R., Cui, F., Townsend, N., Foo, C. C., Guo, J., Shenoi, A., and Xiong, Y. (2016). A review of vibration control methods for marine offshore structures. *Ocean Engineering*, 127, 279–297. https://doi.org/10.1016/j.oceaneng.2016.10.001.

Li, B. H., Gao, H. S., Zhai, H. B., Liu, Y. S., Yue, Z. F., 2011. Free vibration analysis of multi-span pipe conveying fluid with dynamic stiffness method. Nuclear. *Engineering and Design,* 241(3), 666–671. https://doi.org/10.1016/j.nucengdes.2010.12.002.

Liu, L., Shi, K., Fan, X., Tan, W., and Wang, Y. (2021). Risk and characteristics analysis of the flow-induced vibration of the dip tube in opposed multi-burner gasifier. *Journal of Loss Prevention in the Process Industries,* https://doi.org/10.1016/j.jlp.2021.104508.

Liu, Y., Miwa, S., Hibiki, T., Ishii, M., Morita, H., Kondoh, Y., and Tanimoto, K. (2012). Experimental study of internal two-phase flow induced fluctuating force on a 90 elbow. *Chemical Engineering Science*, *76*, 173-187. https://doi.org/10.1016/j.ces.2012.04.021.

Mack, A. ; Joshi, H. ; Belfroid, S. (2018). Numerical rebuilding of dynamic instabilities and forces in multiphase pipe bend flow. *International Journal of Computational Methods and Experimental Measurements*, 6, 358–372. https://doi.org/10.2495/CMEM-V6-N2-358-372.

McNeil, I., & Abdelkefi, A. (2021). Nonlinear modeling and vibration mitigation of combined vortex-induced and base vibrations through energy harvesting absorbers. *Communications in Nonlinear Science and Numerical Simulation*, 95, 105655. https://doi.org/10.1016/j.cnsns.2020.105655.

Miwa, S., Hibiki, T., and Mori, M. (2016). Analysis of flow-induced vibration due to stratified wavy two-phase flow. *Journal of Fluids Engineering*, 138(9). https://doi.org/10.1115/1.4033371.

Miwa, S., Liu, Y., Hibiki, T., Ishii, M., Kondo, Y., Morita, H. and Tanimoto, K.(2014). Study of unsteady gas-liquid two-phase flow induced force fluctuation (Part 1: evaluation and modeling of two-phase flow induced force fluctuation). *Transactions of the JSME* (in Japanese), 80(809), 1-11. https://doi.org/10.1299/transjsme.2014fe0005.

Miwa, S., Mori, M., & Hibiki, T. (2015). Two-phase flow induced vibration in piping systems. *Progress in Nuclear Energy*, 78, 270-284. https://doi.org/10.1016/j.pnucene.2014.10.003.

Mohmmed, A. O., Al-Kayiem, H. H., Nasif, MS, Time, R. W. (2019). Effect of slug flow frequency on the mechanical stress behavior of pipelines. *International Journal of Pressure Vessels and Piping Volume* 172, May 2019, Pages 1-9. https://doi.org/10.1016/j.ijpvp.2019.03.012.

Mohmmed, A. O., Al-Kayiem, H. H., and Osman A. B. (2021). Investigations on the Slug Two-Phase Flow in Horizontal Pipes: Past, Presents, and Future Directives. *Chemical Engineering Science*, 238, 116611 https://doi.org/10.1016/j.ces.2021.116611.

Mohmmed, A. O., Al-Kayiem, H. H., Osman A. B., Sabir, O. (2020). One-way coupled fluid–structure interaction of gas–liquid slug flow in a horizontal pipe: Experiments and simulations. *Journal of Fluids and Structures* 97: 103083. https://doi.org/10.1016/j.jfluidstructs.2020.103083.

Ortiz-Vidal, L. E., Mureithi, N., & Rodriguez, O. (2014). Two- phase friction factor in gas-liquid pipe flow. *Revista de Engenharia Térmica*, 13(2), 81–88. http://dx.doi.org/10.5380/reterm.v13i2.62101.

Ortiz-Vidal, L. E., Rodriguez, O. M. H., and Mureithi, N. W. (2013). An exploratory experimental technique to predict two-phase flow pattern from vibration response. In *Proceedings of Pressure Vessels and Piping Conference*, July 14–18, Paris, France. https://doi.org/10.1115/PVP2013-98115.

Riverin, J. L., Pettigrew, M. J. (2007). Vibration excitation forces due to two-phase flow in piping elements. *Journal of Pressure Vessels Technology*. ASME, 129, 7–13. https://doi.org/10.1115/1.2388994.

Tay, B. L. and Thorpe, R. B. (2004). Effects of liquid physical properties on the forces acting on a pipe bend in gas-liquid slug flow, *Chemical Engineering Research and Design*, 82(3), 344–356. https://doi.org/10.1205/026387604322870453.

Thome, J. R., (2004). Engineering Data Book III, Wolverine Tube, Inc., Huntsville, AL.

Todkar, S. R., Kapale, U. C., Anil, T. R., and Chapgaon, A. N. (2016). An experimental study on two phase (air-water) flow characteristics in a horizontal pipe at atmospheric conditions. *International Journal of Engineering Research & Technology*, *5*, 687-692. https://doi.org/10.17577/IJERTV5IS010553.

Wang, S., and Shoji, M. (2002). Fluctuation characteristics of two-phase flow splitting at a vertical impacting T-junction. *International Journal of Multiphase Flows*, 28(12), 2007–2016. https://doi.org/10.1016/S0301-9322(02)00104-0.

Wongwises, S., Khanhaew, W. & Vetchsupakhun, W., Prediction of liquid hold-up in horizontal stratified two-phase flow. *Thammasat International Journal of Science and Technology*, 3(2), pp. 48–56, 1998.

Wörner, M. (2003). *A compact introduction to the numerical modeling of multiphase flow*. Report in the summer school in Neptun-Olimp, Romania, in the period of July 21 – 25.

Yamagata, K., & Akamatsu, T. (1997). Experimental Study of Flow-Induced Vibration in Two-Phase Flow. *Journal of Fluids Engineering*, 119(3), 622-628. https://doi.org/10.1051/matecconf/201821116001.

Yang, G., Yu, X., Zhang, Y., Chen, G., Bu, S., Zhang, K., and Chen, D. (2024). Numerical analysis on two-phase flow-induced vibrations at different flow regimes in a spiral tube. *Nuclear Engineering and Technology*, 56(5), 1712-1724 https://doi.org/10.1016/j.net.2023.12.025.

Zhu, H., Lin, P., and Yao, J. (2016). An experimental investigation of vortex-induced vibration of a curved flexible pipe in shear flows. *Ocean Engineering*, 121, 62–75. https://doi.org/10.1016/j.oceaneng.2016.05.025.

Chapter 9

Improving the Mechanical Properties of Sandwich Panels with Varying Structures

Salwan H. AlHumairee[1,*]
Ibtihal A. Mahmood[1]
and Mustafa B. Hunain[2]
[1]Mechanical Engineering Department, Faculty of Engineering, University of Technology, Baghdad, Iraq
[2]Mechanical Engineering Department, Faculty of Engineering, University of Babylon, Babylon, Iraq

Abstract

This study explores methods to improve the mechanical properties of composite sandwich panels, including adding polysulfide rubber to epoxy. The optimal concentration depends on the manufacturing method, epoxy properties, and fibers used in the front sheets, enhancing the sandwich panel's strength-to-weight ratio, energy absorption, fire resistance, and insulation. Nanoparticles enhance epoxy and face sheet properties, while cell core shape affects mechanical properties. Honeycomb core sandwich panels are common in aerospace, rail transport, automotive, and medical industries. By changing the shape of the honeycomb cell, the properties of the sandwich panel also change. Different honeycomb topologies include triangle, square, circle, and hexagon geometric shapes, the most common of which is the hexagonal honeycomb core, but in some studies, researchers have optimized the geometric shape of hexagonal cells, which has increased the properties

---

* Corresponding Author's Email: salwanahassan1982@gmail.com

In: Advancements in Sustainability Systems II
Editors: H. H. Al-Kayiem, R. A. Jessam, Hasanain A. A. Wahhab et al.
ISBN: 979-8-89530-562-1

of sandwich panels. The results showed that carbon, Kevlar, glass, and hybrid fibers are the most important fibers for making face sheets, and the mechanical properties of the sandwich panel depend on the arrangement of fibers and the percentage of fibers in the face sheet body. So, by increasing the percentage of Kevlar, glass, and carbon fibers in the face sheet, high-speed impact resistance, flexural strength, and compression strength increased, respectively. The subjects discussed in this study article contain parts of basic sandwich material design, structure design, panel performance, mechanical properties, and failure. Further research directions and potential applications are summarized.

Keywords: sandwich panel, epoxy, fibers, mechanical properties, Nanoparticles, honeycomb

1. Introduction

Sandwich panels are light structures that consist of several parts. Today, these structures have found many uses in various industries due to their properties. The use of sandwich panels in structures can reduce weight and increase the resistance against the forces applied to the structure Kreja, 2011. The use of sandwich panels in different structures, depending on the conditions, can be effective in reducing the construction time (Shen et al. 2010). Today, the ratio of strength or stiffness to weight is one of the most important challenges in the transportation and aviation industries. Sandwich structures are widely favored in contemporary construction for their advantageous combination of lightweight design and robust strength, coupled with their ability to absorb energy efficiently (Rejab et al. 2013).

Sandwich panels typically comprise a core material sandwiched between two face sheets. Depending on the intended application, additional components may be incorporated into the sandwich panels. For instance, in construction, sandwich panels, various types of foams are often introduced within the sandwich structure. These foams may serve a structural purpose or contribute to improving sound and heat insulation properties (Fotsing et al. 2016). Face sheets in sandwich panels can be fabricated from a variety of materials, offering diverse characteristics and functionalities (Hegde et al. 2019). The core of sandwich panels plays a critical role in bearing forces and providing structural integrity. Designing and constructing the core of these panels must be done with careful consideration of the forces they will encounter and various environmental conditions (Ma et al. 2021). The cores of sandwiches are usually a structure with a three-dimensional network, which, due to the increase in the thickness of the structure and its mechanical properties, can have a good performance in bearing the incoming forces (Oliveira et al. 2020). Today, with the development of manufacturing technologies, different types of cores with different properties are made, each of which has its application and sometimes has multiple applications (Xia et al. 2022). Different types of materials with different applications and properties can be used as cores in sandwich panels. The first sandwich panels used in the construction industry had cores made of rock wool and polystyrene. Cores of this type were very welcome due to their insulation against sound and heat transfer and were widely used in the construction of various building components (Proença et al. 2020). Sandwich panels have many uses in the aerospace industry. The cores used in aerospace sandwich panels are usually of the structure called HONEYCOMB, which are usually made of meta-aramid fiber or aluminum sheets (Sun et al. 2021). These types of cores have been used in the aerospace industry for years, but they have weaknesses such as high prices, exclusive and complex manufacturing technology, and the lack of production of different types of cores (Pierre et al. 2023). The existence of weak points of these types of cores in some cases causes reluctance to use these types of cores. Producing cores with fewer weaknesses and especially cores with the possibility of producing different shapes and different geometrical capabilities can make these types of cores superior to conventional cores, and because of this superiority, the use of these types of cores will expand (Naseer, 2021).

In general, the sandwich panel is a light and complex structure with a high tensile, compressive, bending, and twisting load-bearing capacity (Zhang et al. 2021), which is limited to two face sheets on both sides and is placed in the middle of a layer called the core. Face sheets are made of materials with high hardness, strength, and density, such as metal, polymer, composite, and nanocomposite (Chen et al. 2020), while the core is often a material with lower density and hardness compared to the face sheets. The core can be made of foam, honeycomb, and truss, considering that these structures are affected by dynamic and static loadings (O'Hegarty et al. 2020).

Hence, in this review paper, the optimal combination between epoxy and polysulfide, the role of silicon nano carbide in strengthening epoxy, the effect of core shape in improving sandwich panel properties, and the role of the number of face sheet layers were discussed. Also, in this paper, numerical methods have been investigated to estimate the mechanical properties of different sandwich panels. Figure 1 represents a sketch of a honeycomb sandwich panel.

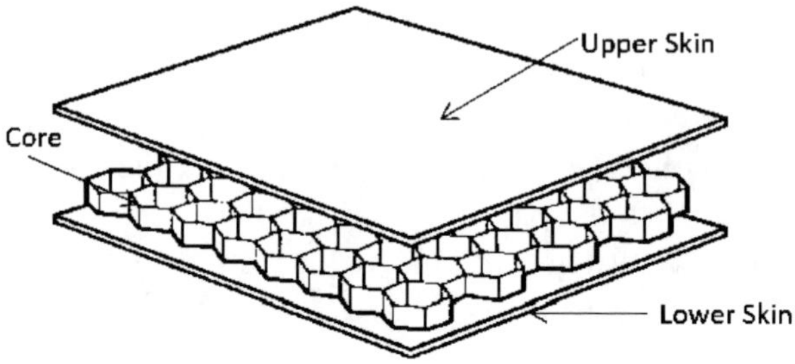

Figure 1. Honeycomb Sandwich Panel.

2. Polysulfide in Improving the Epoxy Properties

Polysulfide is a chemical used as a strengthening and curing agent in epoxy resins (Davoodi et al. 2022). Polysulfide increases the strength and hardness of epoxy resin and also increases its mechanical properties, such as resistance to bending, stretching, and impact (Kader et al. 2020). Another advantage of using polysulfide is improving the thermal resistance of epoxy resin. Polysulfide increases the temperature of the softening point of epoxy resin, which increases the resistance of the resin against high temperatures and its premature softening (Shalaan et al. 2021). Reducing water absorption in epoxy resin is another effect of adding polysulfide to epoxy. Polysulfides reduce water absorption in Epoxy resins and also reduce the relative volume change (Coefficient of Thermal Expansion) at a suitable working temperature of about 100-150 °C (Pirayesh et al. 2019).

Furthermore, polysulfide changes the surface energy of the epoxy resin and also makes the closed cell structure more useful in the epoxy layer, which strengthens the adhesion properties of the epoxy resin (Sobani et al. 2022). In research conducted by Faidzi et al. 2021, the effect of adding epoxy-polysulfide rubber on the mechanical properties of hybrid composite was investigated. The results of this research showed that the use of 2% polysulfide by mass with 98% epoxy improved the tensile strength, impact strength, and hardness of the composite. Kader et al. 2020 investigated the role of polysulfide and Al_2O_3 nanoparticles in improving the properties of Epoxy Polysulfide Copolymer. In this research, 6% polysulfide + 94% epoxy was used for different mass ratios of alumina nanoparticles. The results showed that the addition of alumina nanoparticles improved the tensile strength, bending strength, hardness, and abrasion of the Epoxy-polysulfide copolymer, but with the excessive increase of alumina nanoparticles, these properties decreased. The mechanical properties of (multi-walled carbon nanotubes) MWCNTs-epoxy polysulfide were investigated by Hadavand et al. 2013 (also see Nemaa et al. 2014). They investigated the mechanical properties of epoxy polysulfide resin reinforced with MWCNTs with and without acid functionalization. Three different mass concentrations were defined for MWCNT nanoparticles of 0.1 to 0.3%. The results showed that the most optimal nanoparticle concentration is 0.1% wt. The results showed that the addition of nanoparticles with acid functionalization in epoxy polysulfide resin increased Young's modulus by 57.86%, tensile strength by 66.92%, fracture strain by 56.25%, flexural stress by 63%, and flexural strength by 53%. The mechanical properties of MWCNT/epoxy polysulfide are acquired from many resources and are presented in Table 1.

Table 2. Mechanical properties of MWCNT/epoxy polysulfide

Properties	Unit	MWCNT	pure	0.1 wt.%	0.2 wt.%	0.3 wt.%
Young's Modulus	MPa	Untreated	406	389	478	523
		Acid treated	--	458	535	723
Tensile Strength	MPa	Untreated	4.95	5.24	4.89	4.36
		Acid treated	--	5.29	5.35	8.83
Fracture Strain	%	Untreated	0.24	0.32	0.76	0.27
		Acid treated	--	0.16	0.19	0.25

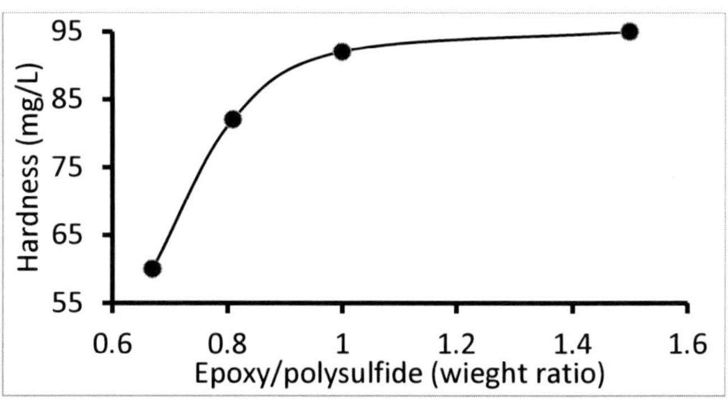

Source: Abdouss et al. 2011.

Figure 2. Effect of adding polysulfide on composite hardness.

Abdouss et al. 2010 investigated the oxidative effects of manganese dioxide and paraquinone dioxide on the properties and curing time of epoxy polysulfide resin. The results of this research showed that oxidative agents had no significant effect on the properties of epoxy polysulfide resin, but using polysulfide in epoxy resin strengthened the mechanical-dynamic properties of the nanocomposite, so that by increasing the mass ratio of epoxy to polysulfide, the hardness of the produced composite increased (Figure 2). By increasing the weight ratio of polysulfide, the tensile strength and hardness of epoxy-polysulfide copolymers decreased, but the percentage of elongation increased.

Polysulfide has an effective role in improving the mechanical properties of epoxy-polysulfide, and by changing its mass concentration, the final properties of epoxy-polysulfide will have significant changes, but choosing the best concentration of polysulfide depends on the final composite. Properties of epoxy, fibers, and reinforcing materials in a sandwich or composite body and the construction method. Figure 3 shows the results presented by Abdouss et al. 2011.

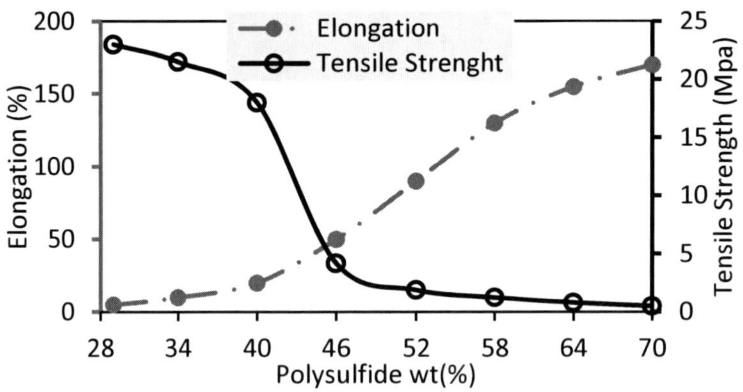

Source: Abdouss et al. 2011.

Figure 3. The effect of polysulfide on the properties of the composite.

3. Nanoparticles' Role in Enhancing Epoxy and Sandwich Panel Properties

The widespread use of epoxy resin has caused much research to investigate the behavior of this material and to modify and improve its characteristics. By examining the IR HNMR (used to study epoxides in lipid oxidation and industrial processes) and CNMR (Nuclear magnetic resonance) spectra of various epoxy resins and investigating the effect of different components such as the type and amount of curing agents and the effect of the presence of fillers and other additives, researchers have started to improve the properties of epoxy resins in such a way that today the role of these resins is it cannot be hidden in different industries (Tarfaoui et al. 2016). Therefore, in this section, the effects of using nanomaterials in improving the properties of epoxy resin have been investigated.

Nanomaterials offer promising opportunities to reduce weight and enhance the performance of structures against various loads, including impact, tension, and bending. Nanoparticles have caused a significant improvement in the properties of composite materials, which is due to the high Young's modulus and extraordinary strength of nanoparticles. Researchers have studied the effect of adding nanoparticles to composite sandwich panels (Sharbaf et al. 2017).

3.1. Silica Nanoparticles

Adding nano-silica improves the properties of pure epoxy (Sharbaf et al. 2017). As an example, Mortazavi et al. 2012 observed that adding nano-silica to epoxy increases the modulus of elasticity, which can be due to the creation of a strong bond between the matrix and nanoparticles. Zheng et al. 2005 investigated the mechanical properties of nanocomposites with glass fibers and nanosilica, which increases the bending strength, shear strength, and compressive strength by increasing the weight percentage of nanosilica. The most appropriate nanoparticle concentration to increase mechanical properties was 3 wt.%. They claimed 115, 13, and 60% increases in the tensile strength, tensile modulus, and impact strength, respectively, with 3 wt.% of nanosilica.

Zamanian et al. 2013 found that adding nano-silica to epoxy increases Young's modulus and fracture energy. With the increase in the diameter of nano-silica, Young's modulus first increased and then decreased, so the highest Young's modulus was observed at a mass concentration of 7 wt.% of nanoparticles with a diameter of 20 nm. However, the maximum fracture energy was observed for nano-silica with a diameter of 12 nm and 6 wt%. One of the possible reasons for this could be the role of nano-silica in preventing crack propagation. In another study, Tsai et al. 2010 investigated the effect of silica nanoparticles on the mechanical properties of composites and reported that by increasing the weight percentage of nano-silica to 40%, fracture toughness, Young's modulus, and shear strength of the composite increased. Wang et al. 2013 found that adding nano-silica to pure epoxy increases Young's modulus and tensile strength. Liu et al. 2011 concluded that the first mode fracture toughness increases with the addition of nanosilica to epoxy, which increases the impact strength by increasing the weight percentage of nanosilica from 0 to 3%. Sing et al. 2021 showed that composites containing 2 wt.% of silica nanoparticles have the maximum tensile strength against impact and stiffness, while the composite with 3 wt.% of silica nanoparticles shows the highest bending strength.

Summing up, the authors produced the mean implications of the added silica nanoparticles to the composite of the sandwich panels, as shown in Figure 4.

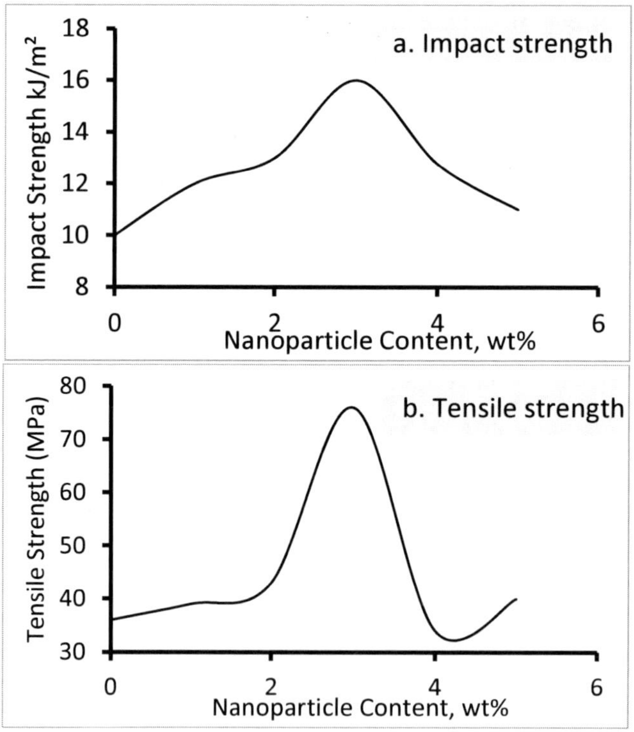

Source: Authors.

Figure 4. Effect of silica nanoparticles on the epoxy nanocomposites in sandwich panels, a. on impact strength, and b. on tensile strength.

3.3. Nanoclay

Nanoclay particles are one of the most important nanometer reinforcements for making nanocomposites. These nanoparticles basically contain silicon, aluminum, or magnesium, oxygen, and hydroxyl with various associated cations. These ions and hydroxyl groups are arranged in sheets in two-dimensional structures (Chan et al. 2011). The structure of nanoclay consists of silicate layers with a thickness of 1 nm, including silicon and aluminum plates, which are connected in different proportions and stacked on top of each other in a certain way with a variable interlayer distance. Among different types of clay minerals, montmorillonite is the most widely used material for preparing polymer nanocomposites reinforced with nanoclay (Jeyakumar et al. 2017). Tirtash et al. 2020 investigated the effects of adding clay nanoparticles to epoxy on the flexural properties of basalt fiber-reinforced composite. The results of this research are shown in Table 2. The results showed that by increasing the concentration of nanoclay up to 5 wt.%, the bending properties of the nanocomposite increased.

Table 3. Three-point bending test results of composite reinforced with clay nanoparticles (Tirtash et al. 2020)

Property	Nano Clay Concentration (w.t%)			
	0	1	3	5
Bending Strength (MPa)	469	520.4	564.8	609.9
Flexural Modulus (GPa)	16.52	18.35	20.98	22.68
Breaking Energy(J/mm^3)	13.49	14.18	17.71	18.94
Breaking Strain	0.047	0.049	0.051	0.054

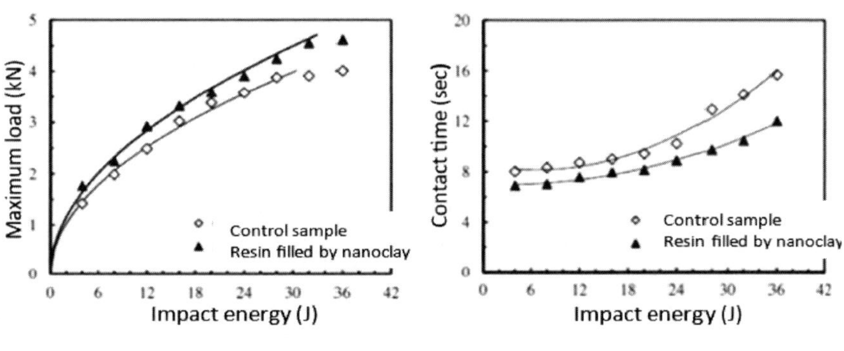

Figure 5. Effect of clay nanoparticles on impact test (Reis et al. 2013).

Joseph et al. 2023 investigated the effect of clay nanoparticles on the mechanical properties of basalt-epoxy face sheets in sandwich panels. The results of this research showed that the use of nanoclay in epoxy increased the bending strength, shear strength of the core, and the amount of energy absorption in the impact test by 34%, 51%, and 72%, respectively.

Reis et al. 2013 investigated the impact resistance properties of sandwich panels reinforced with clay nanoparticles. In Figure 5, the graph of changes in residual flexural strength and maximum load versus impact energy is shown, which shows that for different impact energies, the residual flexural strength and maximum bearing load were always higher in the sandwich panel containing 3 wt.% of nanoclay.

3.3. Nano Silicon Carbide

Silicon Carbide (SiC) is a very hard ceramic material that has a hexagonal or cubic crystal structure. Chemical resistance at high temperatures, semi-conductive properties, high thermal stability, and thermal conductivity have caused it to be considered a practical material for use in electrical equipment that is used at high temperatures. The semiconducting property of SiC can improve the electrical properties of the polymer matrix in polymer-ceramic composites (Rodgers et al. 2005).

Silicon carbide particles help to improve mechanical, electrical, and thermal properties. Epoxy and its composites have common uses, such as coating, underlayment, concrete production, semiconductor encapsulation, and coating (Nassar et al. 2014). Adding silicon carbide to the epoxy structure reduces the shrinkage of epoxy after stress relief and improves its various mechanical properties (Nassar et al. 2013). Integrating the dispersed mineral particles in the cross-sectional area of the polymer ultimately leads to an incredible improvement in the performance of the polymer composites (Nassar et al. 2014). Silicon carbide particles help to improve mechanical, electrical, and thermal properties. Silicon carbide, which is used as a filler in composite structures, greatly increases the thermal conductivity of the composite. Considering this issue, many researchers have used SiC nanoparticles to improve the properties of epoxy and increase the mechanical properties of composites.

Chisholm et al. 2004 at the University of Tuskegee, Alabama, USA, investigated the effect of increasing SiC nanofiller on epoxy resin. The research results showed that the use of 1.5 wt.% and 3 wt.% of SiC nanoparticles improved the mechanical properties of the composite by 20% and 30%. Khashaba et al. 2014 studied the effect of different nanoparticles on the strength of epoxy resin with the aim of modifying epoxy resin. The results showed that nanocomposite materials with 0.5 wt.%.% (MWCNT), 1.5 wt.% of SiC, and 1.5 wt.% of A12O3 have the highest improvement in tensile and shear properties compared to other filler percentages in the desired resin. The tensile strength and tensile modulus of different nano resins increased by 7.5%, 4%, 0.5%, 18.2%, 7.19%, and 1.7%, respectively, compared to the pure resin. Also, shear strength and shear modulus were improved by 5.5%, 4.9%, 6.3%, and 10.3%, 16%, 8.1%, respectively.

Wei et al. 2017 investigated the effect of SiC nanoparticles on the mechanical properties of sandwich panels designed for high-temperature applications. The core of this sandwich panel is a pyramid and is made of ceramic matrix C/SiC composite, which can withstand temperatures up to 1600 °C. The results of the research showed that increasing the temperature caused a decrease in the compressive strength properties of the sandwich panel, so that the compressive strength of the sandwich panel at room temperature, 1200 and 1600 degrees, was equal to 620, 260, and 120 MPa, respectively. In the same conditions and using SiC nanoparticles, the average compressive strength increased by 12.4%, 8.1%, and 5.5%.

Nassar et al. 2013 investigated the effect of the addition of silicon carbide nanoparticles on the mechanical properties of a polymer-based composite. The mass percentage of nano SiC included four levels of 5, 10, 15, and 20 wt.%. The results of this research showed that with increasing SiC concentration in epoxy resin, the wear properties of epoxy resin increased, but the tensile properties of epoxy nano resin decreased compared to pure epoxy resin. Rana et al. 2021 investigated the effect of different concentrations of SiC on the flexural strength of epoxy reinforced with Kevlar fibers. The mass concentrations of nanoparticles included 1 wt.%, 2 wt.%, and 3 wt.% of SiC. The results showed that the use of Kevlar fibers, as well as SiC nanoparticles, increased the flexural strength of the composite. Table 3 shows the results of this research. By increasing the nanoparticle concentration from 1% to 3%, the flexural strength of the composite without Kevlar fibers increased by 81.35% and with Kevlar fibers by 3.40%. Overall, the results showed that the role of SiC nanoparticles in strengthening the pure epoxy phase is much greater than that of fiber-reinforced epoxy.

The research results of Li et al. 2007 showed that due to the provision of a wide surface by SiC nanoparticles, the force between the bonds obtained from epoxy resin is increased, and as a result, its mechanical properties are improved. On the other hand, with the increase in nanoparticle concentration, the nanoparticles inside the epoxy resin become agglomerated, and this problem causes the mechanical properties of epoxy nano resins to decrease. Zheng et al. 2021 stated that silicon carbide nanoparticles are a functional nanoparticle and strongly tend to absorb other nanoparticles; therefore, by increasing the nanoparticle concentration above a certain limit, the nanoparticles strongly attract each other, so coarse agglomeration was produced, which further weakens the mechanical properties of epoxy resin. The properties of reinforced epoxy by nanoparticles depend on the nanoparticles' dispersing method in epoxy, the type of epoxy resin, the type of loading (static or dynamic), the impact energy level, and the type of composite.

Table 4. Three-point bending test results Rana et al. 2021

SiC (wt.%)	Kevlar Fiber	Flexural Strength (MPa)
1	--	34.96
2	--	56.98
3	--	63.4
1	3 Layers	76.3
2	3 Layers	77.6
3	3 Layers	78.9

4. Sandwich Panels of Honeycomb Core

Honeycomb structures have an open-celled arrangement that results from the junction of very thin walls. The cross-section of the cells is usually hexagonal or other geometric shapes. The idea of the honeycomb structure is taken from the nest of honeybees in nature, and more than 500 types of this structure have been made so far. The paper honeycomb structure was made by the Chinese 2000 years ago, which is now the origin of the paper honeycomb structure (Wei et al. 2020). For the first time, a paper honeycomb structure was used for packaging. After that, the honeycomb structure was used in the horizontal beams of the railway(Uddin et al. 2019). The unique feature of honeycomb sandwich structures is their high bending strength and low weight, which has caused increasing attention to these structures. Nowadays, sandwich structures are widely used in many industries, such as aerospace, shipbuilding and boat building, military, automotive, and construction. Currently, the most widely used type of honeycomb is made by using adhesives to connect the sheets that make up the core to each other and the shells (Chandrasekaran et al. 2021).

4.1. Applications

Honeycomb sandwich composites are used in a wide range of engineering sciences due to heat and sound insulation, high strength-to-weight ratio, resistance to compressive loads, and fire resistance (Cahyono et al. 2016). One of the main areas of application of these materials is the construction of structures in aerospace, where honeycomb sandwich panels are used in the construction of rockets, satellite structures, and spacecraft. It is common to use this composite in components such as the flight deck and other interior components of the aircraft (wall, floor, body, roof, toilet storage bin) (Mangino et al. 2007). Mangino et al. 2007 investigated the mechanical and thermal properties of sandwich composites with honeycomb core for use in aircraft bodies. In another research, a sandwich composite with a honeycomb core and carbon fibers was proposed for use in the internal components of the aircraft. Young's modulus, shear modulus, and thermal expansion coefficient of this sandwich panel were obtained as 830 MPa, 4440 MPa, and 2.395 x 10-6/°C, respectively (Chandrasekaran et al. 2021).

Another field of application of honeycomb sandwich composites is the automobile industry. By using these materials, the weight of the car will be significantly reduced. Sukmaji et al. 2017 investigated a honeycomb sandwich panel reinforced with glass and carbon fibers to make the internal components of electric cars. Also, the results of other research showed that the use of honeycomb sandwich panels has a higher level of noise absorption, so they are a suitable option for use in cars.

Another field of application of honeycomb sandwich panels is the high-speed rail transport industry. In this industry, these materials are used to make internal components of high-speed trains (seat molds, internal walls, roof and floor, partitions, etc.). These types of composites are considered because they are light, resistant to impact and pressure, and resistant to corrosive substances (He et al. 2008).

4.2. Honeycomb Topology

The topology and shape of the honeycomb core in the sandwich panel can be found in various forms, such as triangular, square, rectangular, circular, and hexagonal, as outlined in Figure 6. Usually, the most common honeycomb has hexagonal geometry by changing the geometry of the honeycomb core, properties such as strength, stiffness, weight, and ability to make a sandwich panel change (Moghaddam, 2020). Liu and Zhang, 2009 modeled and analyzed the effect of transverse loading on some physical and mechanical properties of sandwich panels with honeycomb cores with triangular and square topologies. The results of their research showed that the density of the sandwich panel with a triangular honeycomb was higher than the sandwich panel with a square honeycomb. Also, the transverse strain created in the sandwich panel under the same loading condition was lower in the sandwich panel with a triangular honeycomb.

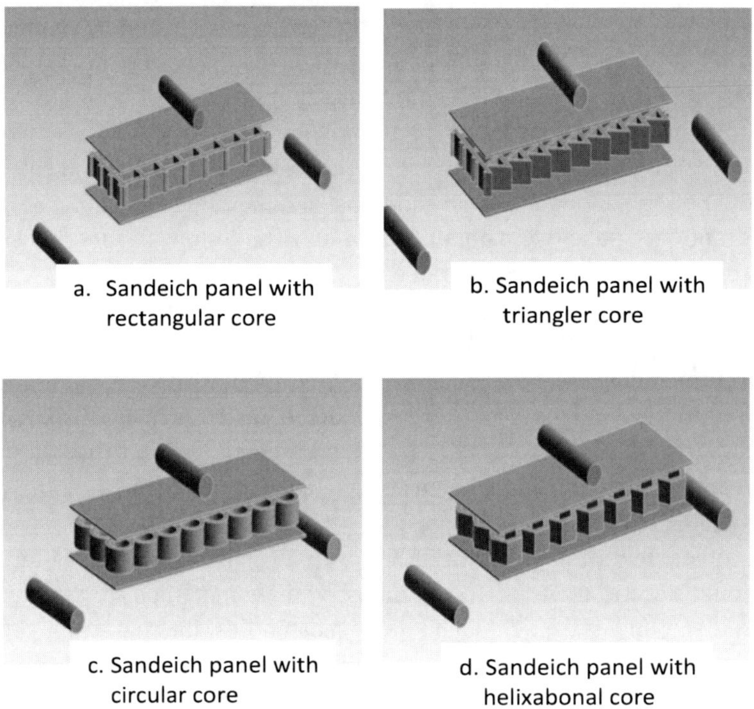

Source: Authors.

Figure 6. Sandwich composite panel with a different core.

Pehlivan et al. 2019 investigated the effect of different honeycomb topologies on the compressive strength of carbon fiber-reinforced polymer (CFRP). Three models, square, circular, and hexagonal, were used for the honeycomb, and the results of this research showed that the thickness of the honeycomb wall influences the CFRP compressive strength of the honeycomb, but its height has no significant effect on compressive strength.

Also, the results showed that a honeycomb with hexagonal topology has a better performance compared to a honeycomb with circular and square topology. Wang et al. 2020 modeled the honeycomb profile with six different topologies using the Lambert-Jonas ballistic principle, and the behavior of each model under the influence of high-speed impact was investigated. The results of the bullet-to-honeycomb impact test with triangular, square, and reentrant cells showed less resistance to the impact, and more honeycomb branches were damaged during the collision. Honeycombs with circular honeycomb in square arrangement (CS type) and circular honeycomb in hexagonal arrangement (CH type) had the highest impact resistance, and they had the greatest effect of absorbing collision energy. The results of the research showed that the ballistic limit speed of the honeycomb with square, triangular, and reentrant cells was lower than the hexagonal model, and on the other hand, the ballistic limit speed of CS and CH models was 15.2% and 25% higher than the hexagonal model, respectively.

In another study, Yap et al. 2015 investigated the effect of compressive loading on three different CFRP models with triangular, circular, and hexagonal honeycombs. The results of this research showed that circular honeycombs have higher strength than hexagonal and triangular honeycombs, but the ultimate compressive strength was the maximum in the hexagonal honeycomb. Examining the results showed that the speed of recovery in triangular honeycombs was higher than in hexagonal and circular honeycombs. Iranmanesh et al. 2023 investigated the effects of seven different honeycomb models with different topologies on panel ballistic properties such as energy absorption capability, amount of absorbed energy, and deflection under high-velocity impact. Different forms of cells of honeycomb profiles are shown to conduct experimental tests. Honeycomb profiles were made by 3D printing. The results of this research showed that honeycombs with Schwartz P cells and tetradial cells have a function compared to other investigated models. According to the results, the best honeycomb cell topology is Schwarz type P, which has higher energy absorption, energy absorption efficiency, and maximum profile deflection in this type than in other models. All the designed models had better performance than the honeycomb with hexagonal cells, except for the Schwartz D model. It can be that there is a difference between FEA and experimental work due to the manufacturing of samples by hand, and the thickness of each layer of the skin cannot take enough of its due amount of epoxy.

Wang et al. 2015 investigated 6 different honeycomb models under ballistic test. These six models included hexagonal, triangular, double hexagonal, circular, full hexagonal, and full circular honeycombs. Different types of honeycombs. The results showed that the maximum energy absorbed by the honeycomb during the impact test and the maximum compressive stress were observed for the honeycomb with a full circular and then circular cell shape. The three studies above showed different results about the best shape of the honeycomb in sandwich panels. The difference in these results is attributed to the difference in materials used, the difference in manufacturing conditions, and the mechanism used in manufacturing, so a difference in the results.

5. Face Sheets in Sandwich Panels

One of the main components of the sandwich panel is the face sheet. Face sheets are usually made of polymer fibers such as Kevlar (Gustin et al. 2005), glass (Rolfe et al. 2018), carbon (Khaledi et al. 2022), and hybrid (Selver et al. 2019) fibers. The process of making face sheets is done by lay-up or vacuum methods, then the created laminate is attached to the sandwich core with glue. Usually, the number of layers of laminate face sheet is variable, which has been investigated by various researchers.

Junaedi et al. 2023 investigated the bending and compressive properties of sandwich panels reinforced with glass fibers. The results presented in this research showed that by changing the percentage of glass fibers in the laminate face sheet, the mechanical properties of the sandwich panel changed. By increasing the percentage of glass in the laminate face sheet from 0 to 30%, the amount of bending force, core shear stress, and stress in the face sheet increased and then decreased so that these properties were maximum at 10 wt.% of glass fiber mass. The compression modulus and compressive strength of the sandwich panel followed a pattern of initial increase followed by a decrease as the percentage of glass fibers in the face sheet increased. Notably, the optimal compressive properties were achieved when the glass fiber content reached 20 weight percent wt.%. Mani et al. 2023 conducted a study on enhancing the mechanical properties of sandwich panels by varying the arrangement of carbon, Kevlar, and glass fibers. They incorporated silica nanoparticles to enhance the bonding between the polymer fibers in the face sheet and the core of the sandwich panel. Their findings revealed that the sandwich panel featuring a face sheet layout of K/C/G/Foam/G/C/K with a nanoparticle concentration of 2 weight percent (w.t%) exhibited the highest tensile, shear, and torsional strength. Selver et al. 2019 investigated eight different sandwich panel models with different face sheets under flexural loading. The manufactured face sheets were a combination of carbon and glass fibers with different rods, which varied according to different forms of carbon and glass rod radius. The sandwich panel with the GC15 face sheet type had the highest flexural and tensile strength.

Rolfe et al. 2018 investigated the effect of the arrangement of carbon and glass fibers in sandwich panels under impact loading. In this research, 5 different models were made for the sandwich panel, including G/G/F/G/G, C/C/F/C/C, G/C/F/C/G, G/C/F/G/C, and G/C/G/F/G/C/G. In situations where the impact surface is known, the use of a hybrid asymmetric face sheet (GCFGC) has a higher impact resistance than other models. The specific deflection and specific strain of this sandwich panel were 0.322 mm.kg/m^2 and 0.077%, but in situations where the impact surface is unknown, it is more appropriate to use a sandwich panel with a symmetrical hybrid face sheet with a higher density. So that the amount of damage to the symmetrical face sheet is more superficial, and the damage to the foam core is less. In this situation, the best sandwich panel has a GCGFGCG arrangement. Specific deflection and a specific strain of sandwich panel with GCGFGCG arrangement after impact were equal to 0.394 mm.kg/m^2 and 0.085%, respectively.

Samlal et al. 2022 investigated the low-speed impact resistance of sandwich panels with different hybrid face sheets (Kevlar (K), carbon(C), and hybrid(H)). In this research, each face sheet consisted of two layers of polymer fibers. In this research, 5 different sandwich panel models were made with different arrangements of fibers. The core of all samples was made of foam, but the arrangement of fibers includes K/K/Foam/K/K (K/K), C/C/Foam/C/Cn (C/C), C/K/Foam/K/C (C/K), K/C/Foam/C/K (K/C), and H/H/Foam/H/H (H/H). The impact test was performed at four levels of 10, 20, 30, and 40 J. The results showed that with the increase in the collision energy level, the amount of damage also increased. Examining the results showed that the maximum damage depth on the surface of the face sheet for C/C, K/K, C/K, K/C, and H/H sandwich panels was equal to 9.45, 3.5, 3.97, 3.95, and 8.75 mm, respectively. The results showed that the highest and lowest resistance to bullet penetration was observed for sandwich panels with K/K and C/C face sheets. In other words, by increasing the number of carbon layers in the face sheet, the penetration resistance of the sandwich panel decreased. The maximum compressive force that can be tolerated by the sandwich panel with face sheets of panels C/C, K/K, C/K, K/C, and H/H was equal to 1368, 3491, 2829, 3723, and 1548 N, respectively. The results of this section also showed that the maximum compressive strength of the sandwich panel was observed in the sandwich panel with a K/C face sheet.

Gustin et al. 2005 investigated the effect of fiber arrangement in the sandwich panel face sheet on the impact properties of the composite. Nine different models were made with carbon and Kevlar fibers to make the laminate face sheet. Each face sheet contains four layers of polymer fibers, which are presented in Table 4. The average absorbed energy and compression strength for sandwich panels with different face sheets are presented in Table 4. The results showed that the lowest and highest amount of absorbed energy was 34.9 and 31 J, which was observed for the sandwich panel with 1: layer Kevlar-3: layer carbon and 4: layer carbon face sheet. Also, the maximum and minimum compression strength of 8.3 and 6 MPa were observed for the sandwich panel with 1: hybrid layer-3: carbon layers and 1: Kevlar-3: carbon layers.

It was found that carbon, Kevlar, and glass fibers are among the most widely used polymer fibers in the construction of face sheet sandwich panels, and by changing the arrangement of fibers in the face sheet, the mechanical properties of the sandwich panel change, so that when the number of carbon layers in the face sheet increases, the high-speed impact resistance of the sandwich panel will decrease. On the other hand, by increasing the number of Kevlar layers in the face sheet, the high-speed impact resistance in the sandwich panel will increase.

Table 5. Effect of fiber arrangement of face sheets on mechanical properties of sandwich panel (Gustin et al. 2005)

Face Sheets arrangement	Average absorbed energy (J)	Compression strength (MPa)
4 Carbon	31	6.5
1 Kevlar- 3 Carbon	34.9	6.0
2 Kevlar-2Carbon	33.2	6.3
3 Kevlar-1Carbon	34.5	7.0
4 Kevlar	33.5	7.7
1 Hybrid-3 Carbon	43.4	8.3
2 Hybrid-2 Carbon	31.9	8.0
3 Hybrid-1 Carbon	33.1	7.9
4 Hybrid	32.8	6.7

The presence of glass fibers in the face sheet improves flexural strength, but increasing the number of carbon fiber layers in the face sheet increases the compression strength. Another result obtained in this section is the effect of symmetric and asymmetric structures in fiber arrangement. In the conditions where the collision surface is known, the asymmetric arrangement of fibers, and in the conditions where the collision surface is unknown, the symmetrical arrangement of fibers has higher energy absorption properties. By increasing the number of face sheet layers, the mechanical properties of the sandwich panel also increase, but this increase in mechanical properties is associated with an increase in weight, density, and physical properties of the sandwich panel, which can have a limiting role in the construction of the sandwich panel. Hence, the determination of the number of face sheet layers is a crucial parameter that should be based on achieving the maximum mechanical properties of the sandwich panel while considering its dimensions.

6. Application of FEM in Analyzing the Mechanical Properties of Sandwich Panels

The Finite Element Method (FEM) serves as a numerical approach utilized for analyzing diverse engineering problems. It is applicable to complex structures of varying shapes, facilitating the analysis of structures with intricate geometric configurations and non-linear behaviors like significant deformations and temperature fluctuations. FEM finds extensive applications across various industries, including aerospace, automotive, construction, plastics, electronics, and more. Recently, numerous researchers have leveraged FEM to scrutinize and explore sandwich and composite structures under diverse loading conditions. Abbadi et al. 2014, employed FEM to simulate the behavior of sandwich panels subjected to fatigue loading. Herranen et al. 2012, utilized FEM to analyze the behavior of Glass Fiber Reinforced Polymer (GFRP) sandwich panels featuring various core materials under flexural loading. The modeling outcomes indicated that the flexural strength of the sandwich panel varied with alterations in the core type. Burlayenko et al. 2009, explored the fatigue characteristics of sandwich panels featuring a honeycomb core using FEM. They employed this approach to ascertain the natural frequencies, load acceptance capacity, and elastic properties of the sandwich panel core. The FEM modeling outcomes revealed that infilling the honeycomb cells with foam led to an augmentation in the buckling stiffness of the sandwich panel. Sakly et al. 2016 employed FE modeling to simulate the ballistic response of sandwich panels utilized in high-speed trains. The outcomes of the FE analysis revealed that during impact testing, the face sheet and core of the sandwich panel experienced separation.

Furthermore, the FE results aligned with the findings of experimental tests, validating the accuracy of the modeling approach.

Table 6. Sandwich panel behavior under different loading conditions

Software	Details	Ref
ABAQUS	circular core honeycomb structure under axial compression behavior, Mild steel was used for making honeycomb panels.	Ghongade 2021
ANSYS	The bending behavior of honeycomb sandwiches made of aluminum alloy cores, four-point bending tests, and the effects of the variation in honeycomb core height and honeycomb sandwich panel skin are studied.	Akiwate et al. 2017
ANSYS	Modeling of dynamic response and energy absorption in an aluminum sandwich panel with a honeycomb core under impact test, the ice wedge impact tests of AHSPs were conducted by using a horizontal impact experimental apparatus.	Wu et al. 2019
ABAQUS	Modeling of low-speed impact behavior of sandwich panel with honeycomb core and CFRP face sheets. The sandwich is made with CFRP skin and Nomex honeycomb core.	Audibert et al. 2019
ABAQUS	Modeling the bending behavior of a sandwich panel with a wooden core in order to determine the energy absorption capacity. The method used to manufacture sandwich panels with prismatic cores is to first veneer sheets of 1.4 mm in thickness, 150 mm in width, and 600 mm in length.	Smardzewski 2019
FEM	Investigating sandwich panel behavior under low-speed impact test and determining compression level by FEM and experimental method with different face sheet thicknesses Composite face sheets are made of unidirectional carbon/epoxy T300/QY8911 prepreg with a ply thickness of 0.125 mm.	Zhang et al. 2016
ABAQUS	Modeling the flexural behavior of sandwich panels with wooden core according to the number of different face sheet layers. Two types of panels are manufactured entirely from wood-based materials, while the third one presents a combination of plywood for surfaces and a corrugated thermoplastic composite as a core part.	Labans et al. 2019
ABAQUS	Investigating the low-speed impact resistance of sandwich panels with different aluminum foam cores and face sheets numerically and experimentally	Huo et al. 2020

Dehuban et al. 2017 studied the effect of fiber orientation and layer arrangement on the buckling behavior of carbon-basalt composite sheets experimentally and numerically. They found that the orientation of the fibers plays a significant role in the compressive load capacity of the composite. The buckling behavior of sandwich panels consisting of honeycomb core and nanocomposite face sheets was studied by FEM (Shahverdi et al. 2019). Zhai et al. 2021, using the numerical analysis of the thermal buckling of the curved sandwich panel, investigated the effect of the panel structure characteristics on the critical buckling temperature (CBT). The bigger the shear modulus of softcore is, the higher the frequencies and corresponding CBTs are. Koleshtri et al. 2022 investigated the buckling of a cylindrical sandwich panel with a functional metal foam core using the FEM. They assumed that the sandwich panel is made of two thin face metal sheets and a moderately thick functionally graded metal foam core. The Navier technique is implemented to derive closed-form solutions for the natural frequencies and the buckling load of the simply supported cylindrical sandwich panels. Results revealed that the presented higher-order porosity distribution could increase the buckling load and fundamental natural frequency by 10% and 5% in comparison to the conventional porosity distributions, respectively. In addition, it is shown that if the flexural vibration is of interest, ignoring the in-plane and rotary inertias, which can significantly reduce the computational complexity, does not have any considerable effect on the natural frequencies.

Table 5 displays the sandwich panel behavior based on FE analysis for different cores and face sheets. So, the numerical results are in good agreement with the experimental results; on the other hand, the examination of the results shows that by using numerical modeling methods, it is possible to reduce the cost of making samples. Also, it is easy to improve the final sandwich panel model by changing the arrangement of face sheet layers, changing the type and geometry of the core, and changing the height of the core and the thickness of the face sheet to improve its mechanical properties.

Conclusion

In this chapter, improved sandwich panel properties were reviewed, and the following results were obtained.

- It can be noted that the mechanical properties of the sandwich panel are significantly enhanced by adding Polysulfide.
- The mechanical properties of the sandwich panel improved by adding nanoparticles due to the high Young's modulus and extraordinary strength of nanoparticles, but with the excessive increase of nanoparticles in epoxy, the nanoparticles became aggregate, causing the creation of microscopic holes in the body of the face sheet or the core of the sandwich panel, finally, it reduces its mechanical properties.

- The results showed that a honeycomb with hexagonal topology has a better performance compared to a honeycomb with circular, square, and triangular topology.
- Carbon, Kevlar, glass, and hybrid fibers are among the most widely used fibers in the face sheet body, and by changing the arrangement of fibers and the number of their layers, the mechanical properties of the sandwich panel changed, such as high-speed impact resistance, flexural strength, and compression strength increased.
- With the FEM, it is possible to model the behavior of the sandwich panel under the influence of different loadings, and by changing the type of core, the shape of the honeycomb cells, the thickness and the type of face sheet, the sandwich panel model can be optimized and thereby reduce costs.

References

Abbadi, A., Tixier, C., Gilgert, J., & Azari, Z. (2015). Experimental study on the fatigue behaviour of honeycomb sandwich panels with artificial defects. *Composite Structures*, 120, 394-405. https://doi.org/10.1016/j.compstruct.2014.10.020.

Abdouss, M., Farajpour, T., and Derakhshani, M. (2010). Investigating of polysulfide and epoxy-polysulfide copolymer curing. *Materialwissenschaft und Werkstofftechnik*, 41(10), 884-888. https://doi.org/10.1002/mawe.201000680.

Abdouss, M., Farajpour, T., and Derakhshani, M. (2011). The effect of epoxy-polysulfide copolymer curing methods on mechanical-dynamical and morphological properties. *Iranian Journal of Chemistry and Chemical Engineering*, 30(4), 37-44. doi: https://doi.org/10.30492/ijcce.2011.6087.

Akiwate, S. B., & Shinde, V. D. (2017). Experimental Investigation of Bending behaviour of Aluminium Alloy Honeycomb Sandwich Structure using Four Point Bending Tests. *International Journal for Innovative Research in Science & Technology*, 4(1), 2349-6010.

Audibert, C., Andréani, A. S., Lainé, É., & Grandidier, J. C. (2019). Discrete modelling of low-velocity impact on Nomex® honeycomb sandwich structures with CFRP skins. *Composite Structures*, 207, 108-118. https://doi.org/10.1016/j.compstruct.2018.09.047.

Burlayenko, V. N., & Sadowski, T. (2009). Analysis of structural performance of sandwich plates with foam-filled aluminum hexagonal honeycomb core. *Computational Materials Science*, 45(3), 658-662. https://doi.org/10.1016/j.commatsci.2008.08.018.

Cahyono, S. I., Widodo, A., Anwar, M., Diharjo, K., Triyono, T., Hapid, A., & Kaleg, S. (2016, March). Lightweight sandwich panel honeycomb core with hybrid carbon-glass fiber composite skin for electric vehicle application. *In AIP Conference Proceedings (Vol. 1717, No. 1). AIP Publishing*. https://doi.org/10.1063/1.4943468.

Chan, M. L., Lau, K. T., Wong, T. T., Ho, M. P., & Hui, D. (2011). Mechanism of reinforcement in a nanoclay/polymer composite. *Composites Part B: Engineering*, 42(6), 1708-1712. https://doi.org/10.1016/j.compositesb.2011.03.011.

Chandrasekaran, N. K., & Arunachalam, V. (2021). State-of-the-art review on honeycomb sandwich composite structures with an emphasis on filler materials. *Polymer Composites*, 42(10), 5011-5020. https://doi.org/10.1002/pc.26252.

Chen, G., Cheng, Y., Zhang, P., Liu, J., Chen, C., Zhao, Y., and Wang, H. (2020). Contact underwater explosion response of metallic sandwich panels with different face-sheet configurations and core materials. *Thin-Walled Structures*, 157, 107126. https://doi.org/10.1016/j.tws.2020.107126.

Chisholm, N., Mahfuzi, H., Rangari, V., Rodgers, R., & Jeelani, S. (2004). Synthesis and mechanical characterization of carbon/epoxy composites reinforced with SiC nano particles. *NSTI-Nanotech*, 3, 302-307.

Davoodi, F., Akbari-Kharaji, E., Danaee, I., Zaarei, D., & Shishesaz, M. (2022). Corrosion behavior of epoxy/polysulfide coatings incorporated with nano-CeO2 particles on low carbon steel substrate. *Corrosion*, 78(8), 785-798. https://doi.org/10.5006/4116.

Faidzi, M. K., Abdullah, S., Abdullah, M. F., Azman, A. H., Hui, D., & Singh, S. S. K. (2021). Review of current trends for metal-based sandwich panel: Failure mechanisms and their contribution factors. *Engineering Failure Analysis*, 123, 105302. https://doi.org/10.1016/j.engfailanal.2021.105302.

Fotsing, E. A., Leclerc, C., Sola, M., Ross, A., & Ruiz, E. (2016). Mechanical properties of composite sandwich structures with core or face sheet discontinuities. *Composites Part B: Engineering*, 88, 229-239. https://doi.org/10.1016/j.compositesb.2015.10.037.

Ghongade, G., Kalyan, K. P., Vignesh, R. V., & Govindaraju, M. (2021). Design, fabrication, and analysis of cost effective steel honeycomb structures. *Materials Today: Proceedings*, 46, 4520-4526. https://doi.org/10.1016/j.matpr.2020.09.694.

Gustin, J., Joneson, A., Mahinfalah, M., & Stone, J. (2005). Low velocity impact of combination Kevlar/carbon fiber sandwich composites. *Composite structures*, 69(4), 396-406. https://doi.org/10.1016/j.compstruct.2004.07.020.

Hadavand, B. S., Javid, K. M., & Gharagozlou, M. (2013). Mechanical properties of multi-walled carbon nanotube/epoxy polysulfide nanocomposite. *Materials & Design*, 50, 62-67. https://doi.org/10.1016/j.matdes.2013.02.039.

He, M., & Hu, W. (2008). A study on composite honeycomb sandwich panel structure. *Materials & Design*, 29(3), 709-713. https://doi.org/10.1016/j.matdes.2007.03.003.

Hegde, S. R., & Hojjati, M. (2019). Effect of core and facesheet thickness on mechanical property of composite sandwich structures subjected to thermal fatigue. *International Journal of Fatigue*, 127, 16-24. https://doi.org/10.1016/j.ijfatigue.2019.05.031.

Herranen, H., Pabut, O., Eerme, M., Majak, J., Pohlak, M., Jaan, K. E. R. S., ... & Aruniit, A. (2012). Design and testing of sandwich structures with different core materials. *Materials Science*, 18(1), 45-50. https://doi.org/10.5755/j01.ms.18.1.1340.

Huo, X., Liu, H., Luo, Q., Sun, G., & Li, Q. (2020). On low-velocity impact response of foam-core sandwich panels. *International Journal of Mechanical Sciences*, 181, 105681. https://doi.org/10.1016/j.ijmecsci.2020.105681.

Iranmanesh, N., Sarvestani, H. Y., Ashrafi, B., & Hojjati, M. (2023). Beyond honeycombs: Core topology's role in 3D-printed sandwich panels. *Materials Today Communications*, 37, 107548. https://doi.org/10.1016/j.mtcomm.2023.107548.

Jeyakumar, R., Sampath, P. S., Ramamoorthi, R., & Ramakrishnan, T. (2017). Structural, morphological and mechanical behaviour of glass fibre reinforced epoxy nanoclay composites. *The International Journal of Advanced Manufacturing Technology*, 93, 527-535.

Jiang, H., Liu, X., Jiang, S., & Ren, Y. (2023). Hybrid effects and interactive failure mechanisms of hybrid fiber composites under flexural loading: Carbon/Kevlar, carbon/glass, carbon/glass/Kevlar. *Aerospace Science and Technology*, 133, 108105. https://doi.org/10.1016/j.ast.2023.108105.

Joseph, C., Muthukumar, C., Ng, L. F., Subramanian, J., Ramesh, C., Krishnasamy, S., & Thiagamani, S. M. K. (2023). The effect of nanoclay on the performance of basalt-epoxy facesheet and foam core sandwich panels. *Journal of Sandwich Structures & Materials*, 25(7), 730-746. https://doi.org/10.1177/10996362231191140.

Junaedi, H., Khan, T., & Sebaey, T. A. (2023). Characteristics of carbon-fiber-reinforced polymer face sheet and glass-fiber-reinforced rigid polyurethane foam sandwich structures under flexural and compression tests. *Materials*, 16(14), 5101. https://doi.org/10.3390/ma16145101.

Kader, E. E., Abed, A. M., & Al-Shammari, M. A. (2020). Al2O3 Reinforcement Effect on Structural Properties of Epoxy Polysulfide Copolymer. *Journal of Mechanical Engineering Research and Developments*, 43(4), 320-328.

Khaledi, H., & Rostamiyan, Y. (2022). Experimental Investigation of High-velocity Impact Effects on Composite Sandwich Panel with M-Type Lattice Core Made of Carbon Fiber Reinforced by Nano-SiO2. *Journal of Solid and Fluid Mechanics*, 11(6), 127-142. https://doi.org/10.22044/jsfm.2021.10445.3323.

Khashaba, U. A., Aljinaidi, A. A., & Hamed, M. A. (2014). Nanofillers modification of Epocast 50-A1/946 epoxy for bonded joints. *Chinese Journal of Aeronautics*, 27(5), 1288-1300. https://doi.org/10.1016/j.cja.2014.08.007.

Kreja, I. (2011). A literature review on computational models for laminated composite and sandwich panels. *Open Engineering*, 1(1), 59-80. https://doi.org/10.2478/s13531-011-0005-x.

Labans, E., Kalnins, K., & Bisagni, C. (2019). Flexural behavior of sandwich panels with cellular wood, plywood stiffener/foam and thermoplastic composite core. *Journal of Sandwich Structures & Materials*, 21(2), 784-805. https://doi.org/10.1177/109963 6217699587.

Li, H., Zhang, Z., Ma, X., Hu, M., Wang, X., & Fan, P. (2007). Synthesis and characterization of epoxy resin modified with nano-SiO2 and γ-glycidoxypropyltrimethoxy silane. *Surface and Coatings Technology*, 201(9-11), 5269-5272. https://doi.org/10.1016/j.surfcoat.2006.07.143.

Liu, H. Y., Wang, G. T., Mai, Y. W., & Zeng, Y. (2011). On fracture toughness of nanoparticle modified epoxy. *Composites Part B: Engineering*, 42(8), 2170-2175. https://doi.org/10.1016/j.compositesb.2011.05.014.

Liu, Y., and Zhang, X. C. (2009). The influence of cell micro-topology on the in-plane dynamic crushing of honeycombs. *International Journal of Impact Engineering*, 36(1), 98-109. https://doi.org/10.1016/j.ijimpeng.2008.03.001.

Ma, Q., Rejab, M. R. M., Siregar, J. P., & Guan, Z. (2021). A review of the recent trends on core structures and impact response of sandwich panels. *Journal of Composite Materials*, 55(18), 2513-2555. https://doi.org/10.1177/0021998321990734.

Mangino, E., Carruthers, J., & Pitarresi, G. (2007). The future use of structural composite materials in the automotive industry. *International Journal of vehicle design*, 44(3-4), 211-232. https://doi.org/10.1504/IJVD.2007.013640.

Mani, M., Thiyagu, M., & Krishnan, P. K. (2023). Experimental investigation of Kevlar/carbon/glass/polyurethane foam epoxy hybrid sandwich composites with nano silicon particles in low-velocity impact events. *Materials Today: Proceedings*. https://doi.org/10.1016/j.matpr.2023.03.209.

Moghaddam, H. S. (2020). *A numerical homogenization approach to characterize in-plane anisotropic hyperelastic responses of a non-metallic honeycomb core*. Doctoral dissertation, Wichita State University.

Mortazavi, B., Bardon, J., Bomfim, J. A. S., & Ahzi, S. (2012). A statistical approach for the evaluation of mechanical properties of silica/epoxy nanocomposite: Verification by experiments. *Computational Materials Science*, 59, 108-113. https://doi.org/10.1016/j.commatsci.2012.03.002.

Naseer, Z., and Khan, Z. (2021). Graphene effect on mechanical properties of sandwich panel for aerospace structures. *Key Engineering Materials*, 875, 121-126. https://doi.org/10.4028/www.scientific.net/KEM.875.121.

Nassar, A., and Nassar, E. (2014). Thermo and mechanical properties of fine silicon carbide/chopped carbon fiber reinforced epoxy composites. *Universal Journal of Mechanical Engineering*, 2(9), 287-292. https://doi.org/10.13189/ujme.2014.020903.

Nassar, A., and Nassar, E. (2013). Study on mechanical properties of epoxy polymer reinforced with NanoSiC particles. *Nanoscience and Nanoengineering*, 1(2), 89-93. https://doi.org/10.13189/nn.2013.010201.

Nemaa, A., Al-Razaq, I A., and Kader, E. E. (2014). Characteristics and Properties of Epoxy/Polysulfide Composite Materials Reinforced by Carbon Nanotubes. *Diyala Journal of Engineering Sciences*, 07(3), 106-119. https://doi.org/10.24237/djes.2014.07307.

O'Hegarty, R., & Kinnane, O. (2020). Review of precast concrete sandwich panels and their innovations. *Construction and building materials*, 233, 117145. https://doi.org/10.1016/j.conbuildmat.2019.117145.

Oliveira, P. R., May, M., Panzera, T. H., Scarpa, F., & Hiermaier, S. (2020). Improved sustainable sandwich panels based on bottle caps core. *Composites Part B: Engineering*, 199, 108165. https://doi.org/10.1016/j.compositesb.2020.108165.

Pehlivan, L., & Baykasoğlu, C. (2019). An experimental study on the compressive response of CFRP honeycombs with various cell configurations. *Composites Part B: Engineering*, 162, 653-661. https://doi.org/10.1016/j.compositesb.2019.01.044.

Pierre, J., Iervolino, F., Farahani, R. D., Piccirelli, N., Lévesque, M., & Therriault, D. (2023). Material extrusion additive manufacturing of multifunctional sandwich panels with load-bearing and acoustic capabilities for aerospace applications. *Additive Manufacturing*, 61, 103344. https://doi.org/10.1016/j.addma.2022.103344.

Pirayesh, A., Salami-Kalajahi, M., Roghani-Mamaqani, H., & Najafi, F. (2019).

Polysulfide polymers: synthesis, blending, nanocomposites, and applications. *Polymer Reviews*, 59(1), 124-148. https://doi.org/10.1080/15583724.2018.1492616.

Proença, M., e Sousa, A. N., Garrido, M., & Correia, J. R. (2020). Acoustic performance of composite sandwich panels for building floors: Experimental tests and numerical-analytical simulation. *Journal of Building Engineering*, 32, 101751. https://doi.org/10.1016/j.jobe.2020.101751.

Rana, R. S., Buddi, T., & Purohit, R. (2021). Effect of SiC reinforcement on the mechanical properties of Kevlar fiber based hybrid epoxy composites. *Materials Today: Proceedings*, 44, 2478-2481. https://doi.org/10.1016/j.matpr.2020.12.542.

Reis, P. N. B., Santos, P., Ferreira, J. A. M., & Richardson, M. O. W. (2013). Impact response of sandwich composites with nano-enhanced epoxy resin. *Journal of Reinforced Plastics and Composites*, 32(12), 898-906. https://doi.org/10.1177/0731684413478993.

Rejab, M. R. M., & Cantwell, W. J. (2013). The mechanical behaviour of corrugated-core sandwich panels. *Composites Part B: Engineering*, 47, 267-277. https://doi.org/10.1016/j.compositesb.2012.10.031.

Rodgers, R. M., Mahfuz, H., Rangari, V. K., Chisholm, N., & Jeelani, S. (2005). Infusion of SiC nanoparticles into SC-15 epoxy: an investigation of thermal and mechanical response. *Macromolecular Materials and Engineering*, 290(5), 423-429. https://doi.org/10.1002/mame.200400202.

Rolfe, E., Kaboglu, C., Quinn, R., Hooper, P. A., Arora, H., & Dear, J. P. (2018). High velocity impact and blast loading of composite sandwich panels with novel carbon and glass construction. Journal of Dynamic Behavior of Materials, 4, 359-372.

Sakly, A., Laksimi, A., Kebir, H., & Benmedakhen, S. (2016). Experimental and modelling study of low velocity impacts on composite sandwich structures for railway applications. *Engineering Failure Analysis*, 68, 22-31. https://doi.org/10.1016/j.engfailanal.2016.03.001.

Samlal, S., & Santhanakrishnan, R. (2022). Low-velocity impact behavior of foam core sandwich panels with inter-ply and intra-ply carbon/kevlar/epoxy hybrid face sheets. *Polymers*, 14(5), 1060. https://doi.org/10.3390/polym14051060.

Selver, E., & Kaya, G. (2019). Flexural properties of sandwich composite laminates reinforced with glass and carbon Z-pins. *Journal of Composite Materials*, 53(10), 1347-1359. https://doi.org/10.1177/0021998318800146.

Shahverdi, H., Barati, M. R., & Hakimelahi, B. (2019). Post-buckling analysis of honeycomb core sandwich panels with geometrical imperfection and graphene reinforced nanocomposite face sheets. *Materials Research Express*, 6(9), 095017. DOI https://doi.org/10.1088/2053-1591/ab2b74.

Shalaan, M. M., AL-Falahi, A. H., & AL-Dabagh, B. M. (2021, March). Study of mechanical properties and thermal conductivity of polymeric blend [epoxy and polysulfide rubber (EP + PSR)] reinforced by nano ceramic powder ZrO2. *IOP Journal of Physics, Conference Series* 1829 012013 https://doi.org/10.1088/1742-6596/1829/1/012013.

Sharbaf, H. R., Rahimi, G. H., & Liaghat, G. H. (2017). Experimental study of behavior of filament winding composite pipes with liner using glass fibers and silica nanoparticles under impact loading. *Journal of Science and Technology of Composites*, 3(4), 311-320.

Shen, J., Lu, G., Wang, Z., & Zhao, L. (2010). Experiments on curved sandwich panels under blast loading. *International Journal of Impact Engineering*, 37(9), 960-970. https://doi.org/10.1016/j.ijimpeng.2010.03.002.

Singh, T., Gangil, B., Ranakoti, L., & Joshi, A. (2021). Effect of silica nanoparticles on physical, mechanical, and wear properties of natural fiber reinforced polymer composites. *Polymer Composites*, 42(5), 2396-2407. https://doi.org/10.1002/pc.25986.

Smardzewski, J. (2019). Wooden sandwich panels with prismatic core–Energy absorbing capabilities. *Composite Structures*, 230, 111535. https://doi.org/10.1016/j.compstruct.2019.111535.

Sobani, M., & Soucek, M. D. (2022). Low temperature fracture toughness of polysulfide modified BPA-epoxide primers. *Progress in Organic Coatings*, 163, 106626. https://doi.org/10.1016/j.porgcoat.2021.106626.

Sukmaji, I. C., Wijang, W. R., Andri, S., Bambang, K., & Teguh, T. (2017, January). Application of sandwich honeycomb carbon/glass fiber-honeycomb composite in the floor component of electric car. In *AIP Conference Proceedings* 1788(1). AIP Publishing. https://doi.org/10.1063/1.4968309.

Sun, G., Huo, X., Wang, H., Hazell, P. J., & Li, Q. (2021). On the structural parameters of honeycomb-core sandwich panels against low-velocity impact. *Composites Part B: Engineering*, 216, 108881. https://doi.org/10.1016/j.compositesb.2021.108881.

Tarfaoui, M., Lafdi, K., & El Moumen, A. (2016). Mechanical properties of carbon nanotubes based polymer composites. *Composites Part B: Engineering*, 103, 113-121. https://doi.org/10.1016/j.compositesb.2016.08.016.

Tirtash, B., Montazeri, A., Eslami-Farsani, A., Bagheri Tirtash, A., Montazeri, A., & Eslami-Farsani, R. (2020). An investigation on the flexural performance of basalt fibers-epoxy hybrid composites reinforced with nanoclay particles. *Journal of Science and Technology of Composites*, 6(4), 525-532. https://doi.org/10.22068/jstc.2020.32816.

Tsai, J. L., Hsiao, H., & Cheng, Y. L. (2010). Investigating mechanical behaviors of silica nanoparticle reinforced composites. *Journal of composite materials*, 44(4), 505-524. https://doi.org/10.1177/0021998309346138.

Uddin, M. N., Gandy, H. T., Rahman, M. M., & Asmatulu, R. (2019). Adhesiveless honeycomb sandwich structures of prepreg carbon fiber composites for primary structural applications. *Advanced Composites and Hybrid Materials*, 2, 339-350.

Wang, J., Waas, A. M., & Wang, H. (2013). Experimental and numerical study on the low-velocity impact behavior of foam-core sandwich panels. *Composite Structures*, 96, 298-311. https://doi.org/10.1016/j.compstruct.2012.09.002.

Wang, Y., Yu, Y., Wang, C., Zhou, G., Karamoozian, A., & Zhao, W. (2020). On the out-of-plane ballistic performances of hexagonal, reentrant, square, triangular and circular honeycomb panels. *International Journal of Mechanical Sciences*, 173, 105402. https://doi.org/10.1016/j.ijmecsci.2019.105402.

Wang, Z., Zhang, Y., and Liu, J. (2015, August). Comparison between five typical reinforced honeycomb structures. In 2015 *International Conference on Advanced Engineering Materials and Technology* (pp. 704-707). Atlantis Press. https://doi.org/10.2991/icaemt-15.2015.134.

Wei, X., Xiong, J., Wang, J., and Xu, W. (2020). New advances in fiber-reinforced composite honeycomb materials. *Science China Technological Sciences*, 63(8), 1348-1370. https://doi.org/10.1007/s11431-020-1650-9.

Wu, X., Li, Y., Cai, W., Guo, K., & Zhu, L. (2022). Dynamic responses and energy absorption of sandwich panel with aluminium honeycomb core under ice wedge impact. *International Journal of Impact Engineering*, 162, 104137. https://doi.org/10.1016/j.ijimpeng.2021.104137.

Xia, F., Durandet, Y., Tan, P. J., & Ruan, D. (2022). Three-point bending performance of sandwich panels with various types of cores. *Thin-Walled Structures*, 179, 109723. https://doi.org/10.1016/j.tws.2022.109723.

Yap, Y. L., & Yeong, W. Y. (2015). Shape recovery effect of 3D printed polymeric honeycomb: This paper studies the elastic behaviour of different honeycomb structures produced by PolyJet technology. *Virtual and Physical Prototyping*, 10(2), 91-99. https://doi.org/10.1080/17452759.2015.1060350.

Zamanian, M., Mortezaei, M., Salehnia, B., & Jam, J. E. (2013). Fracture toughness of epoxy polymer modified with nanosilica particles: Particle size effect. *Engineering Fracture Mechanics*, 97, 193-206. https://doi.org/10.1016/j.engfracmech.2012.10.027.

Zhai, Y., Ma, J., Yan, Y., Li, Q., Wang, S., and Wang, G. (2021). Thermal buckling and free vibration of composite sandwich curved panels. *Composite Structures*, 267, 113869. https://doi.org/10.1016/j.compstruct.2021.113869.

Zhang, T., Yan, Y., Li, J., & Luo, H. (2016). Low-velocity impact of honeycomb sandwich composite plates. *Journal of Reinforced Plastics and Composites*, 35(1), 8-32. https://doi.org/10.1177/0731684415612573.

Zhang, Y., Li, Y., Guo, K., & Zhu, L. (2021). Dynamic mechanical behaviour and energy absorption of aluminium honeycomb sandwich panels under repeated impact loads. *Ocean Engineering*, 219, 108344. https://doi.org/10.1016/j.oceaneng.2020.108344.

Zheng, J., Zhang, X., Cao, J., Chen, R., Aziz, T., Fan, H., & Bittencourt, C. (2021). Behavior of epoxy resin filled with nano-SiO2 treated with a Eugenol epoxy silane. *Journal of Applied Polymer Science*, 138(14), 50138. https://doi.org/10.1002/app.50138.

Zheng, Y., Ning, R., & Zheng, Y. (2005). Study of SiO2 nanoparticles on the improved performance of epoxy and fiber composites. *Journal of reinforced Plastics and Composites*, 24(3), 223-233. https://doi.org/10.1177/0731684405043552.

Chapter 10

The Investment of Nanotechnology in Enhancing Thermal Energy Storage for Solar Energy Technology: A Review

Mohammed Y. Nawaf[1]
Abdulrazzak Akroot[1]
and Hasanain A. Abdul Wahhab[2,*]

[1]Department of Mechanical Engineering, Faculty of Engineering, Karabük University, Karabük, Turkey
[2]Training and Workshop Center, University of Technology, Baghdad, Iraq

Abstract

As a potential contender for attaining a significant boost in thermal storage performance, nanotechnology-based solar energy is receiving much interest. The need for increasing solar system efficiency in solar-related sectors is expanding despite the fact that such solar energy is anticipated to be more expensive. When the breakeven point for high performance is reached, and its cost becomes competitive with other energy sources, it is necessary to dominate the energy portion. The efficiency of solar systems has been targeted through a variety of strategies. Nanotechnology applications enable more cost-effective solar gadget manufacturing. Effective solar systems for cost-effective conversion, efficient solar energy storage systems, or solar energy on a big scale can all be produced using nanofluids to increase efficiency. In

* Corresponding Author's Email: 20085@uotechnology.edu.iq

In: Advancements in Sustainability Systems II
Editors: H. H. Al-Kayiem, R. A. Jessam, Hasanain A. A. Wahhab et al.
ISBN: 979-8-89530-562-1

order to improve solar systems using nanofluids and nanoparticles, this study looks at some of the most recent initiatives and pressing issues.

Keywords: solar energy, thermal storage energy, phase change material, solar collector system, nanotechnology

1. Introduction

When looking for thermal energy storage (TES), substances meeting the following requirements are preferred: low heat loss while storage, a large energy extraction effectiveness, a suitable temperature, non-hazardousness to the surrounding environment, it is available in a way that allows it to be used continuously, considering the cost-effectiveness compared to the rest (Zalba et al. 2003, Abhat, 1983). The reviews of TES for Flat-Plate Solar Collector (FPSC) applications are explored in this study. In general, there is a disparity between the rate and timing of solar energy collection and the load requirements of the thermal application due to the intermittent, variable, and unexpected character of solar irradiation (Dincer & Rosen, 2002; Lane, 1983). Therefore, using a storage system is frequently necessary and required. In general, energy can be stored when the amount gathered is greater than what is needed for the application, and energy can be released when the amount collected is insufficient. Because of its improved load control, energy storage (ES) technologies are widely recognized for their contribution to an energy system. By shifting peak period high demand to off-peak low demand, peak reduction increases generation capacity (Dincer & Rosen, 2002). Through reducing thermal and electricity-generating units, the operation of power plants and thermal machinery at nominal capacity, the use of cheaper off-peak energy, and the reduction of fossil fuel-based output, TES achieves benefits for the economy and the environment (Ayyappan et al. 2015).

The present study is a review of progress in the field of the FPSC systems integrated with TES, which can eliminate the disadvantage of solar energy use for water heating beyond solar mode hours. Phase change Materials (PCMs) are commonly used as TES materials in solar collector systems due to their high storage capacity (Karkri et al. 2015; Xiao et al. 2013). The most widely used organic PCM is paraffin wax, mostly due to its availability, chemically inertness, self-nucleating, small phase change segregation, and non-corrosive (Aldoss & Rahman, 2014). However, the low thermal conductivity and volumetric storage density exhibited are their limitations. These limitations can be overcome with the addition of nanomaterials. Conventional working fluids, such as water, have low thermal conductivity and are subjected to freeze under lower temperatures (Karthikeyan et al. 2014). The use of nanofluid as a working fluid can increase the rate of energy absorption and release, thereby enhancing the overall system performance.

In revealing the exhibited thermo-physical properties of nanoparticles in base fluid and composite, this present study emphasizes the utilization of nanocomposite as a source of energy storage material while the nanofluid is used as working fluid in the developed nano-additives FPSC integrated with TES systems for SWH to provide continuous hot water production beyond solar mode hours (Alkilani et al. 2011). The research questions in this thesis are explored from a bottom-up approach (Zhou & Zhao, 2011). Thus, the design configuration for the collector, the material composition for the TES and working fluids, the coil heat exchanger configuration, and the thermal storage tank stratification for the whole system will be explored.

2. Background on Thermal Energy Storage

Solar thermal energy could be stored as sensible heat, latent heat, thermochemical, or a combination of these, which results from the change in internal energy of the storage material (Hasnain, 1998). It is an effective means by which excess heat collected during periods of sunshine can be preserved for future use in situations of poor solar radiation supply, cloudy days, and at night (Sorour, 1988). The need for thermal energy storage is mainly for effective and efficient energy conservation and management. This is because there is a mismatch between the solar thermal energy generated during the day and the instantaneous demand. Similarly, solar energy is an intermittent energy source that undergoes fluctuation in supply as a result of atmospheric conditions such as cloudiness, dust, and rain. To elaborate on the above point, the design of solar collectors considers the mean solar radiation intensity at a location, which is always less than the maximum solar radiation intensity generated near solar noon. Near solar noon, for a solar air heater, the heat generated leads to high thermal losses to the ambient. To reduce these thermal losses, the need for heat storage is important. The important characteristics of a heat storage medium are cost-effectiveness and delivery of the stored heat to its application points at a desirable temperature (Afolabi et al. 2017). Thermal energy storage can be based on the process of energy gain and storage, as shown in Figure 1.

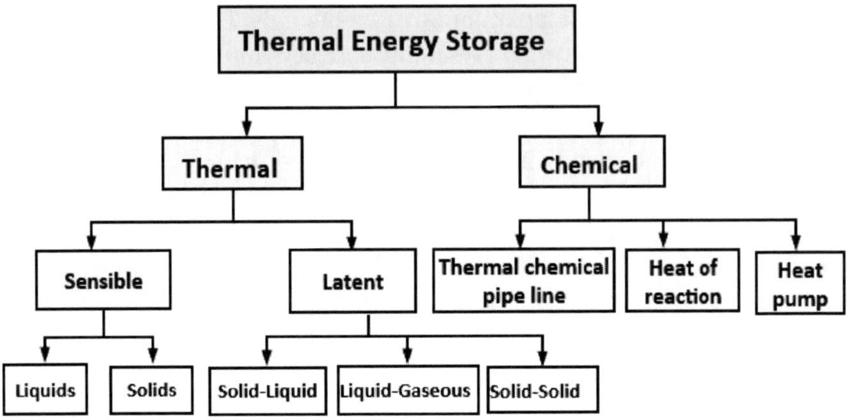

Source: Afolabi et al. 2017 with permission.

Figure 1. Solar Thermal Storage System Overview.

2.1. Sensible Heat Storage- SHS

Sensible heat storage systems have been widely used because of the cost benefits and the availability of materials. Thermal energy storage by sensible heat is achieved by increasing the temperature of the storage material (solid or liquid), which has high heat capacity. During charging and discharging, the storage material utilizes the heat capacity and temperature change to its advantage (Sharma et al. 2009). The amount of thermal energy stored by a sensible heat storage medium is a function of the mass of the material, the heat capacity of the material, and the change in temperature within the material, which can be presented mathematically using Equation 1 (Zhang et al. 2016).

$$Q_{sen-store} = m\int_{T_i}^{T_f} c_p \Delta T = mc_p\left(T_f - T_i\right) \tag{1}$$

where m is the mass of storage material, Cp is the SHS material's specific heat at constant pressure, Ti is the initial temperature of the material, and T_f is the final temperature of the material used for air heating applications, as in the case of the solar chimney collector. Zhang et al. 2016 reported that rock bed-type storage materials are more suitable. Considering the various applications where heat storage is necessary, it is important to identify the advantages and disadvantages of various thermal storage materials. The SHS materials' advantages and disadvantages are described in Table 1.

Table 1. Advantages and disadvantages of SHS systems

Advantages	Disadvantages
The cost of the storage materials is cheap	Low heat storage capacity per unit volume of the storage medium
The materials are locally available, like water, stones, salty water, etc.	Non-isothermal behavior during heat storage (charging) and heat release (discharging) processes
High specific heat capacity	The energy cannot be stored or released at a constant temperature.

2.2. Latent Heat Storage- LHS

Thermal energy storage by latent heat storage is based on heat absorption or discharge when a thermal storage material undergoes a phase change. The latent heat storage materials are otherwise PCMs, which are divided into three categories, namely: organic compounds, inorganic compounds, and mixed compounds (eutectics). The heat storage capacity of PCM can be determined from Equation 2.

$$Q_{lat-store} = \int_{T_i}^{T_m} mc_{ps}\Delta T + m\Delta h_m + \int_{T_m}^{T_f} mc_{pl}\Delta T \quad (2)$$

where T_m is the melting point of the phase change material, Cp_s is the specific heat of the PCM in the solid state, Cp_l is the specific heat of the PCM in the liquid state, and Δh_m is the phase change enthalpy of the material.

In situations where the Cp_s and Cp_l are not dependent on temperature, the heat storage capability can be calculated using Equation 3.

$$Q_{lat-store} = m\left(c_{ps}(T_m - T_i) + \Delta h_m + c_{pl}(T_f - T_m)\right) \quad (3)$$

2.3. Thermochemical Energy Storage

This energy storage system depends on the energy absorbed and released by chemical bond fission and fusion in a completely reversible chemical reaction. The amount of heat stored is a function of the mass of the mass of the storage material, m, endothermic heat of reaction, h_r, and thermal conversion rate of the material, α_r, which can be mathematically represented using Equation 4.

$$Q_{Them-store} = m\alpha_r \Delta h_r \tag{4}$$

with the challenges of space requirement, latent heat thermal storage is particularly attractive because of the associated high energy storage density. LHS materials also store and discharge heat at a constant temperature, unlike the sensible heat storage system. PCM stores 5-14 times more energy per unit volume than sensible heat materials. On the other hand, the sensible heat storage system is cheap, and in a situation where space requirement is not a constraint, it can be a better substitute for the LHS system.

3. Phase Change Materials

PCM is capable of storing and releasing large amounts of energy as a result of the high heat of fusion during phase transition at certain temperatures (Al-Abidi et al. 2012). The PCM can be categorized into three broad groups, namely: a) Organic, b) Inorganic, and c) Eutectics. They are available in different temperature ranges for specific use (Sharma et al. 2009). Some of the desirable properties of PCM are the small volume change, high storage density per unit volume, low vapor pressure, chemical stability, non-corrosiveness and non-toxicity, non-flammability and non-segregation, self-nucleation, adequate thermal conductivity, low cost, and availability as highlighted in the literature (Agyenim et al. 2010, Al-Abidi et al. 2013). In spite of the desirable properties exhibited by PCM, most time, they do not sufficiently meet the requirements of energy storage methods and, as such, require a quantity to overcome the limitation (Ermis et al. 2007). For instance, using metallic fins to improve PCMs' thermal conductivity. Multi-PCMs are used to improve heat storage. By incorporating a nucleating agent into the storage medium, super-cooling can be reduced, and incompatible melting can be prevented by using sufficient thickness (Gil et al. 2010; Mat et al. 2013). Paraffin and non-paraffin compounds are the most often used organic PCMs for the storage of thermal energy applications. The organic materials have self-nucleation, which means they crystallize with little to no super-cooling, and congruent melting, which means they melt and freeze again without phase separation, resulting in the loss of their latent heat of fusion. They are also non-corrosive.

Fatty acids and other compounds are non-paraffins. Non-paraffin materials, which have unique, distinctive features and are thought to be the biggest category of materials for phase change storage, contrast with the paraffin sub-group. Fatty acids, glycols, and alcohol are some of them. Non-paraffin materials have a number of limitations, including the fact that they are flammable and cannot be subjected to intense heat, flames, or oxidizing chemicals (Mondal, 2008).

Inorganic PCMs can also be divided into metals and salt hydrates. Inorganic salts and water combine to form salt hydrates. They are frequently employed because of their excellent thermal conductivity and high fusion heat. They are often affordable enough to be used in thermal energy storage applications and also have minor volume changes when changing from solid to liquid form. However, they frequently display sub-cooling or cooling below the melting point without crystallization, and many inorganics are corrosive. This is because they frequently have weak nucleating properties. In addition, they frequently melt inconsistently or partially, which causes segregation and an irreversible chilling or heating of the salt.

Two or more PCMs that melt and freeze simultaneously make eutectics. The three types of eutectics are organic-organic, inorganic-organic, and inorganic-inorganic. Eutectics have little to no phase segregation issues. Because of their high volumetric heat storage density and narrow melting temperature ranges, eutectics are also regarded as good PCMs (Regin et al. 2008).

3.1. Paraffin

Paraffin wax is a member of straight chain n-alkanes, CH_3-(CH_2)-CH_3. Generally, the longer the length of the pure hydrocarbon chain is, the greater the melting temperature and specific heat of fusion, density, and molar specific heat capacity (of liquid and solid) (Regin et al. 2008; Ukrainczyk et al. 2010). The availability of paraffin in a large temperature range enables its use as a heat of fusion storage material. Since technical grade paraffin is inexpensive and non-corrosive, it is frequently utilized as PCM in latent heat storage devices. They exhibit little volume changes while melting, are chemically inert and steady under 500°C, and have low vapor pressure when in the melt state. System-using paraffin typically has a lengthy freeze-melt cycle because of these characteristics of the paraffins. Paraffins melt congruently and are self-nucleating; however, they exhibit low thermal conductivity, incompatibility with plastic containers, and considerable flammability are some unfavorable characteristics. The wax and the storage unit can be slightly modified to prevent all these negative effects. The thermo-physical properties of some technical grade Paraffins have been reported by many researchers, e.g., Erkizia et al. 2024, Kim et al. 2020, and She et al. 2019. Table 2 describes the features of the classes of PCMs and the main advantages and disadvantages.

Table 2. Features of several types of PCMs

Authors	Classification	Advantages	Disadvantages
Erkizia et al. 2024 Ayyappan et al. 2015 She et al. 2019	Organic PCMs	1. Availability in a large range 2. High heat of fusion 3. No Super-cooling 4. Chemically stable and recyclable 5. Good compatibility with other materials	1. Low thermal conductivity (0.2 W/m·K) 2. Relatively large volume change 3. Flammability
Kim et al. 2020 Tian et al. 2022	Inorganic PCMs	1. High heat of fusion 2. High thermal conductivity around (0.5 W/m·K) 3. Low volume change 4. Availability of low-cost	1. Super-cooling 2. Corrosion
Yeon, 2020 Pilehvar et al. 2019	Eutectics	1. Sharp melting temperature 2. High volumetric thermal storage density	Lack of currently available test data on thermo-physical properties.

3.2. Enhancement Techniques in PCM Thermal Storage

Several studies have been conducted to investigate heat transfer enhancement techniques in PCMs (Robak et al. 2011; Camacho et al. 2012). Finned tubes with various designs (Colella et al. 2011), bubble agitation, inserting a metal matrix into the PCM, employing PCM dispersed with high conductivity particles (Parry et al. 2014; Xie et al. 2013), and micro-encapsulating of the PCM are some of the approaches used, and shell-in-tube or multi-tube (Rahimi et al. 2014; Fleming et al. 2015). The review of latent heat thermal energy storage systems was carried out by Agyenim et al. 2010, outlining the several PCMs examined over the past three decades, the phase change issue formulation, and the heat transfer and enhancement methods used in PCMs to efficiently charge and discharge latent heat energy.

Latent heat thermal energy storage (LHTES) using heat pipes was experimentally examined by Robak et al. 2011. The PCM energy storage quantities, solidification, melting rates, and photographic observations were reported. We looked at the heat transfer properties of technical grade paraffin (RT100) utilized as a phase transition material in a shell-and-tube LHTES.

According to a paper by Colella et al. 2011, a 15% graphite fraction by volume has been proven to be effective.

To determine how fin and copper foam improve the heat transfer performance of a PCM TES, Xi et al. 2013 conducted an experimental investigation. According to the stated experimental findings, using fin and copper foam can cause a sample to melt significantly more quickly than using only copper foam, and the FCF composite has an effective thermal conductivity of 10.83 W/m·K, which is 2.7 times greater than that of the CFP composite. Additionally, for high thermal conductivity composites, horizontal placement contributes more to heat transmission than vertical placement.

Rahimi et al. 2014 examined the melting and solidification of paraffin RT35 in a fined tube to ascertain the enhanced potential of utilizing fins and finless heat tube designs. As a result of the findings, melting and solidification processes are improved during phase changes. Fleming et al. (2015) looked into how adding open-cell aluminum foam improved heat transmission in a shell-and-tube latent heat thermal storage unit utilizing water as the PCM. According to the overall heat transfer coefficient, the testing results showed that the foam considerably boosted heat transfer, while in both solidification and melting, there have been several enhancement techniques have been studied to possibly improve the collector performance over time.

In spite of the improvement achieved using some of the applied techniques, the performance of the collector is not actually improved by all enhancement strategies, according to research by numerous authors.

4. Enchantment Heat Transfer Using Additive Techniques

Conventional methods of enhancement have come with limitations of either settling the macro/micron elements in the composite or an increase in the size of the PCMs containers and extra weight. Increasing conductive heat transfer ratios in the PCMs by inserting high thermal conductivity carbon fiber brushes into the shell side of a heat exchanger is a common additive enhancement studied by many researchers. The additives of nano-sized powders, fillers, fibers, and graphite flakes are useful to increase the thermal conductivity of PCMs. However, the percentage fraction of additives needed to disperse into PCMs to meet certain applications has still been understudied by many researchers.

Paraffin PCM is enhanced with Metal nanoparticles and used in energy storage. Nanoparticles of metals and metal oxides are synthesized with PCMs of different base compositions to increase thermal conductivity and performance.

4.1. Additives

Zhou & Zhao 2011 embedded paraffin wax and calcium chloride hex hydrate into Cu metal foam and expanded graphite. Results showed that Cu metal foam provided a better heat transfer rate as compared to expanded graphite because of continuous interconnected structures. The melting time of paraffin-filled copper metal foam was reduced from 10 minutes to 8 minutes. In contrast, calcium chloride hexahydrate-filled copper metal foam was reduced from 16 minutes to 12 minutes. For the case of expanded graphite, the melting time of paraffin was reduced from 18 minutes to 15 minutes, and calcium chloride was reduced from 28 minutes to 20 minutes, respectively.

Wang et al. 2012 mixed micro-silica or silica fume (SF) with stearic acid (SA) at a 46:56 percentage weight loading ratio. Results showed that the sub-cooling effect was reduced for the PCM composite. This was observed in the melting time and solidification time of silica fume mixed with stearic acid (SF/SA), which was reduced by 33% and 65.6% when compared with SA, respectively. Thus, indicating that SF enhanced the heat transfer rate of SA to form a PCM composite for TES applications.

A numerical simulation of hybrid PCM with graphene fillers for Li-ion batteries was conducted by Goli et al. 2014. The study demonstrates that adopting hybrid graphene fillers over traditional PCMs alone can significantly improve the thermal management and reliability of Li-ion batteries. An increase in the percentage composition of graphene fillers increases the thermal capability of Li-ion batteries. However, an optimized peak of 20% percentage fraction was observed to give the best result.

The expansion and shrinking of PCM, in particular the Paraffin wax, is a challenge for the user as it causes a leakage. One of the mitigation methods is the encapsulation in containers. Al-Kayiem and Alhamdo, 2012 performed an experimental and numerical study on a Paraffin wax encapsulated in a packed bed. The results enhanced the understanding of the melting and solidification during the charging and discharging process. Yu et al. 2014 created PCM microcapsules with an n-octadecane core and a Calcium Carbonate ($CaCO_3$) shell using a self-assembly technique to improve thermal conductivity and durability. The microencapsulated H-octadecane demonstrated good phase-change performance and attained a high heat storage capability. Thermogravimetric research demonstrated that the $CaCO_3$ shell's microencapsulation of octadecane increased the evaporation temperature of octadecane, which in turn improved thermal stability. Due to the creation of a highly thermally conductive $CaCO_3$ shell, the thermal conductivity of octadecane microcapsules was greatly increased.

The outcomes of Chen et al. 2014 experimental investigation of the PCM melting behavior in pore-scaled metal foams demonstrate that metal foam can improve solid-liquid phase-change heat transfer due to thermal conduction in the metal matrix. To simulate the two-dimensional melting of PCM in metal foams, a thermal lattice Boltzmann model with doubled populations was used. A novel method was created by Vitorino et al. 2014 to process self-assembled core-shell composites for thermal control or heat storage. Numerical results using a core-shell model demonstrate that even without external encapsulation, the shape stabilization and microstructure of the graphite shell remain stable during repeated cycling above and below the paraffin melting temperature.

4.2. Nanocomposites

Nano additives of metal and metal-size particles are classified as nanoparticles. Nanotubes of carbon nanotubes (CNT), single wall carbon nanotubes (SWCNT), and multiwall carbon nanotubes (MWCNT) are most common. However, the pointers on the development and characterization of the nanocomposites for TES in conjunction with solar collectors are the research team at Universiti Teknologi PETRONAS, headed by Al-Kayiem. The team reported their development and testing results of Cu-Paraffin nanocomposite in publications, e.g., Lin and Al-Kayiem, 2012; Al-Kayiem and Lin, 2014; and Owolabi et al. 2014.

Experimental measurements of paraffin nanocomposite of CuO and TiCh are studied by Harikrishnan et al. 2014 with Sodium dodecyl benzenesulfonate (SDBS) as a surfactant. The characteristic behavior of Nanocomposites with respect to phase transition and thermal reliability of nanocomposites after several cycles was reported. The result showed thermal conductivity increase with an increase in mass fraction of Nano-additives of CuO in a linear manner, while for TiCh increase is non-linear and inconsistency in thermal conductivity increments. The TiCh Nano composite enhanced paraffin base material higher than CuO by all measured increments by mass fraction of nano-additives. In all, experimental results proved that the heating and cooling rates of the composite PCMs nanocomposite were faster due to the dispersion of nano additives.

The TES of paraffin-based composite PCMs packed with hexagonal boron nitride (h-BN) nanosheets without a surfactant was experimentally evaluated by Fang et al. 2014 at varied percentages by weight composition. It was demonstrated that increasing the loading of h-BN nanosheets increased the thermal conductivity of composite PCMs by a factor of up to 60% in both the solid and liquid phases. When h-BN nanosheets are added to composite PCMs, it is discovered that their latent heat of fusion marginally lowers with loading while their melting/solidification points are essentially unaffected. In addition, it was evident that the improved thermal conductivity of the composite PCMs enabled speeding up of both melting and solidification heat transfer rates.

Nano-octadecane was examined as being enclosed in a St/MMA copolymer shell by Tumurah et al. (2014). A one-step, narrow particle size distribution mini-emulsion in situ polymerization procedure was used to prepare the synthesis. The St/MMA mass ratio significantly influences the effectiveness of n-octadecane encapsulation and the form of nanocapsules. According to TGA/DTG data, the St/MMA copolymer shell has some degree of protection for n-octadecane. After 360 heating/cooling cycles, nanocapsules demonstrated good thermal reliability in terms of chemical stability and thermal characteristics. The obtained nano-encapsulated PCM (Sample A4) has a good potential to be employed for thermal energy storage applications, such as building applications, with phase change enthalpies of 107.9 J (melting) at 29.5°C and 104.9 J (crystallization) at 24.6°C.

Sahan & Paksoy, 2014 investigation the effects of incorporating nanomaterials on improving paraffin's thermal performance for TES systems. Magnetite nano-additives were introduced to paraffin without surfactant in weight fractions of 1%, 5%, 10%, and 15%. Thermal analysis using DSC reveals that while the melting temperature range stayed constant, the LHS capacity of paraffin for micromagnetic composite rose by 10%. Given that magnetite has a thermal conductivity 50 times greater than that of paraffin, a notable rise in the thermal conductivity of the paraffin nanocomposite is also anticipated.

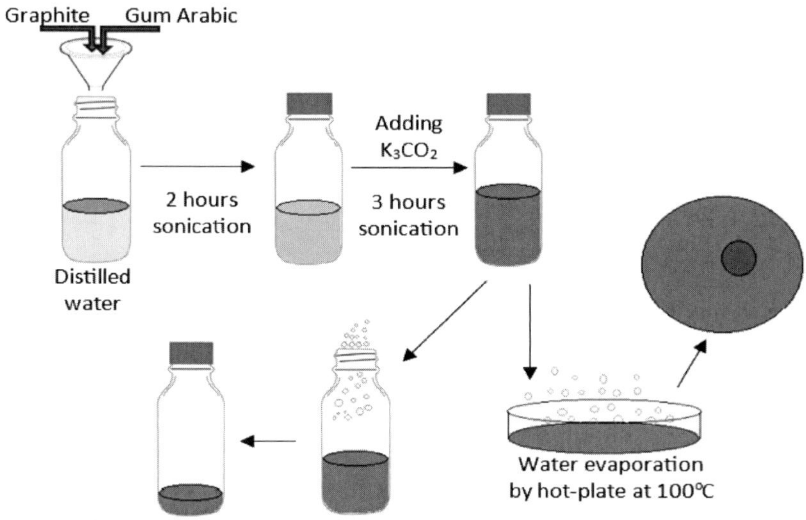

Source: Authors.

Figure 2. Schematic of the synthesis methods for molten salts Nanomaterials.

In combinations of carbonate salts doped with graphite nanoparticles, Jo & Banerjee 2014 investigated the impact of nanoparticle dispersion on specific heat capacity. The results showed that adding 0.1 wt.% of graphite nanoparticles has greatly increased the specific heat capacity of carbonate salt mixes in both the solid and liquid phases. In the solid and liquid phases, respectively, the maximum increases in heat capacity for nanomaterials are 40% and 57%. Figure 2 shows a schematic representation of the synthesis process for molten salt and nanomaterials.

Harish et al. 2015 examine temperature-dependent thermal conductivity increase in single-walled carbon nanohorn inclusions implanted in molten lauric acid phase change nanocomposite. Experimental results revealed differential increases in the thermal conductivity of nanocomposites in the solid and liquid phases, with the greatest increases in the solid and liquid phases at 2% volume fraction being found to be 37% and 11%, respectively.

In an innovative design, PCM and solar collector were combined in a single device, compacting the solar water heater and the nanocomposite TES. Two techniques have been reported to enhance the charging and discharging of heat from/to the Cu-Paraffin nanocomposite underneath the absorber plate of the collector. Figure 3a shows the compacted system with extended plates from underneath the absorber and immersed in the nanocomposite container.

Figure 3b shows another method of increasing the contact area between the backside of the absorber plate and the nanocomposite using aluminum foam.

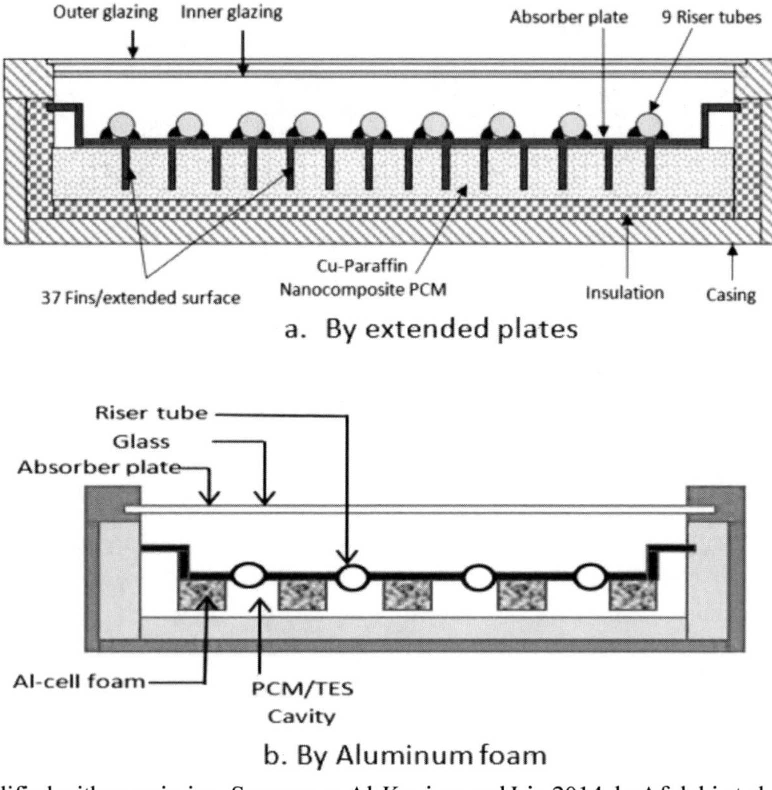

Modified with permission. Sources: a: Al-Kayiem and Lin 2014, b: Afolabi et al. 2016.

Figure 3. Illustrations of heat transfer enhancement in compacted solar water heater integrated with nanocomposite PCM TES.

4.3. Nanofluids

Metallic and nonmetallic nanometer-sized particles with distinctive thermal, magnetic, electrical, optical, and mechanical properties are produced using nanotechnology. According to studies, nanofluids in solar systems have distinct benefits over traditional fluids because of their extraordinary capacity for heat transfer. Raja et al. (2013) conducted an experimental examination using nanofluid in a convective heat transfer in a horizontal circular pipe with fluid in a mixed laminar flow range. Al_2O_3 nanofluid with a water basis is used. Results show that using nanofluids increases the heat transfer coefficient in simple tubes by 8–12% compared to the flow of water in a plain tube. When compared to water, a nanofluid with a 0.5% particle concentration had the most friction. Increases in particle concentration led to increases in the Nusselt number and friction factor. However, when the flow's Reynolds number grows, friction is reduced, but the Nusselt number rises.

For direct absorption in solar collector systems, Luo et al. 2014 evaluated exotherm oil production using various nano-additives of carbon nanotubes, TiO_2, Al_2O_3, Ag, Cu, SiO_2, and graphite utilizing 2-step techniques. According to the findings, the conductivity of nanofluid decreased as the temperature rose, much like it did in the matrix, and increased as nanoparticle loadings increased. Low loadings can effectively increase the absorption of 200–2000 nm radiation. The radioactive characteristics of nanofluids were related to layer depth. Experimental findings were consistent with predictions made by the theoretical model.

In comparison to basic oil, nanofluids increased exit temperature and efficiency by 30-100 K and 2-25%, respectively. The performance of nanoparticles (C, Ag, SiO_2, Al_2O_3, and Cu) to increase DAC efficiency is satisfactory, with the exception of TiO_2. Saha & Paul, 2014, turbulent forced convection flow of water-based Al_2O_3 and TiO_2 nanofluids flowing through a horizontal circular pipe with a uniform heat flux boundary condition applied to the wall has been numerically investigated.

When compared to TiO_2-water nanofluid, a KO_3-water nanofluid exhibits a greater rate of heat transfer. Additionally, it was discussed how the size, diameter, and Brownian movement of nanoparticles have a greater impact on heat transfer performance than the thermal conductivity of nanofluid. Dawood et al. (2014) reported on numerical simulations of laminar mixed convection heat transfer with nanofluids of Al_2O_3, CuO, SiO_2, and ZnO as working fluids in an elliptic annulus with uniform heat flux. The results showed that SiO_2 with a working fluid concentration of 4%, particle diameter of 20 nm, and Reynolds number of 1000 provided the highest heat transfer improvement through an elliptic annulus. The Nusselt number is noticeably higher when the nanoparticle volume percentage and Reynolds number rise, but it falls as the nanoparticle diameter rises. The results show that water-SiO_2 produces the lowest Nusselt number, followed by glycerine-SiO_2 and engine oil-SiO_2, and EG-SiO_2, respectively.

The friction factor and convective heat transfer coefficient for fully developed turbulent flow of MWCNT-Fe_3O_4/water hybrid nanofluids moving through a uniformly heated, constant heat flux circular tube were studied by Sundar et al. (2014). In comparison to base fluid data, the results show a maximum increase in Nusselt number of 31.10% with a penalty of 1.18 times more pumping power. This is for particle loading of 0.3% at a Reynolds number of 22,000. At a base fluid temperature of 60 °C, a 0.3% particle concentration increases viscosity by 1.5 times and thermal conductivity by 29%. The enhancement in the Nu number for 0.1% of MWCNT-Fe304 hybrid nanofluid is 9.35%, 20.62%, for 0.3% of MWCNTFe304 hybrid nanofluid is 14.81%, and 31.10%, respectively, compared to the base fluid.

Ray et al. 2014 developed a mathematical model to examine the performance of three nanofluids: aluminum oxide, copper oxide, and silicon dioxide nanoparticles in a small mini-channel plate heat exchanger. According to the theoretical findings, three different nanofluids with 1.0% volume fractions each transported roughly 2.5 kW of heat, and all of them outperformed the base fluid in terms of efficiency. The convective heat transfer coefficient was improved by 4.75% using aluminum oxide at a flow rate of 0.2 kg/s on the hot side of PHE; when compared to the base fluid, EG/W, copper oxide reduced the volumetric flow rate by 4.78% and needed pumping power by 1.73%. While the most encouraging results on the basis of heat dissipation of 2.5 kW were pumping power savings: 4.72% for Al_2O_3 and 3.01% for CuO. At a pumping power of 0.586 W, the Al_2O_3 nanofluid provided the best heat transfer surface area reduction of about 0.86%.

Bandana et al. 2014 conducted an experimental evaluation of the photo-thermal conversion characteristics and performance of plasmatic silver nanofluids in direct absorption for the solar collector. The findings demonstrate that silver particles, even at very low concentrations, have outstanding photo-thermal conversion capabilities. Peak temperature results in up to a 144% increase in thermal energy storage for a particle concentration of 6.5 ppm. At 6.5 ppm, the specific absorption rate is almost constant; however, as concentrations rise, it sharply declines due to an increase in particle-particle interactions.

4. Key Findings

- Due to their remarkable mechanical, electrical, optical, magnetic, and surface properties, nanomaterials have garnered much attention. The high surface-to-volume ratio of these materials has important effects on energy storage (Yu et al. 2014; Dheep & Sreekumar, 2014). The main characteristics of this novel class of materials for thermal energy storage devices are their high surface area and potential for nanomaterial consolidation (Mai et al. 2014). Nanoscale materials can be advantageous if specific physical and chemical properties associated with the nanoscale can be effectively utilized (Pielichowska & Pielichowski, 2014; Al-Kayiem et al. 2013). The current work focuses on the usage of nano-additives in FPSCs coupled with TES systems to improve paraffin base PCMs and mixtures of water/glycol base fluid (Assaf et al. 2023; Alesbe et al. 2023). According to Figure 4, nanomaterials can be categorized as organic and inorganic materials (Ajel et al. 2023). CNT, SWCNT, MWCNTs, graphite, and nanofibers are all examples of organic nanomaterials (Khafaji et al. 2022).

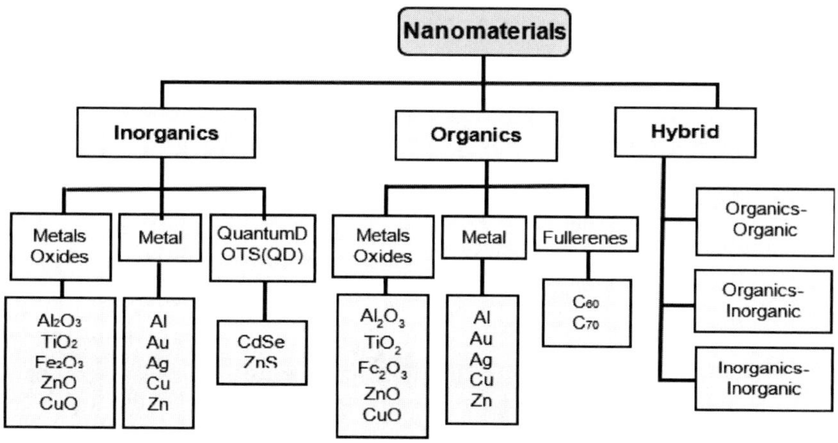

Source: Authors.

Figure 4. Summary of classification of Nanomaterials.

- Phase change materials, such as nanocomposites, are often used for thermal energy storage applications where low-temperature operation is required (Sukhatme & Sukhatme, 1996; Tiwari, 2002). Paraffin wax is a polymer matrix-type composite because of its long-chain carbon-carbon properties (Zalba et al. 2003). When paraffin wax is mixed with Fe nanoparticles, it forms polymer matrix Nanocomposites (PMC) or polymer Nanocomposites (Madodi et al. 2023; El-Adawy et al. 2023). Matrix means the dominant material in the composite, and it can be polymer, metal, or ceramic in this category (Al-Maliki et al. 2022; Sarkar et al. 2015). Figure 5 demonstrates various matrix composite breakdowns.

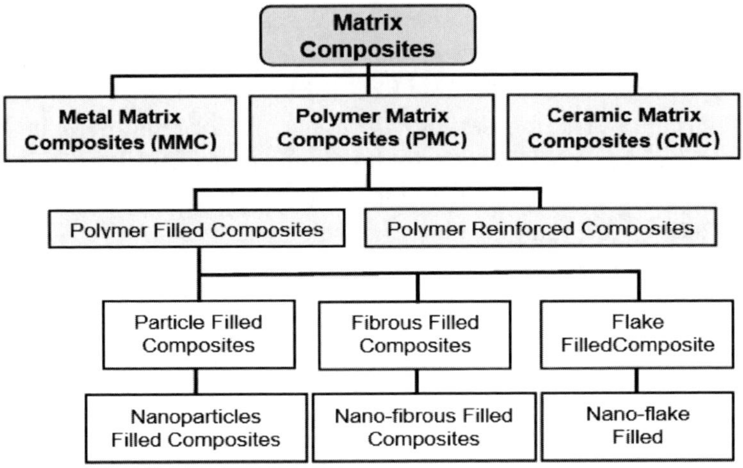

Source: Authors.

Figure 5. Polymer Nanocomposites chart.

- In comparison to basic fluids like water and oil, nanofluids have been discovered to have improved thermal conductivity, thermal diffusivity, viscosity, and convective heat transfer coefficients. It has shown a wide range of possible applications (Sidik et al. 2014; Teng et al. 2014). Some significant difficulties must be stressed for a two-phase system. One of the most important issues is the stability of nanofluids. The review articles involving the progress of nanofluid investigation were published by several researchers (Murshed & de Castro, 2014; Gupta et al. 2014). The thermophysical properties and convective heat transfer of nanofluids have been reviewed experimentally and theoretically, with few studies on their application to various technology applications. The present work focuses on the use of nanofluids for TES in solar water heating. The summary description of the hybrid TES mediums by nano additives is shown in Figure 6.

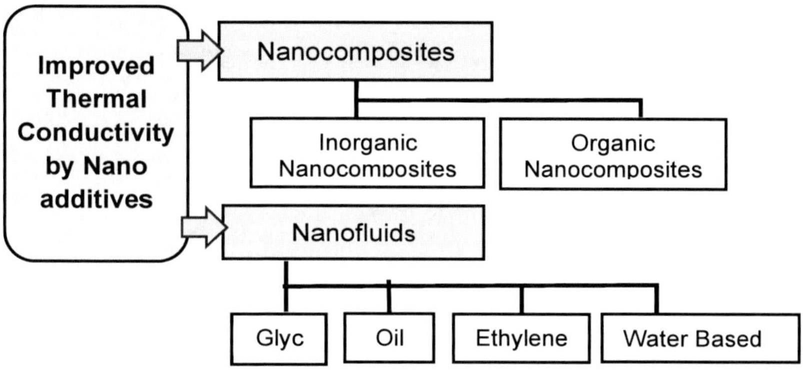

Source: Authors.

Figure 6. Illustration of different nanoenhanced base fluids and composites.

Conclusion

The solar system is one of the most well-known sources of renewable energy. The use of conventional solar collectors for heating, cooling, and heating of both indoor and outdoor spaces is widespread and a well-established technique. The absorption characteristics of the working fluid, which are quite poor for a typical traditional solar flat plate collector, however, limit the thermal efficiency of these collectors. The thermal efficiency of solar flat plate collectors has recently been discovered to increase by 35% when nanofluids, which are essentially liquid-nanoparticle colloidal dispersions as a working fluid, are used. The last twenty years have seen the emergence of sophisticated fluids called nanofluids, which include nanoscale particles. Many solar energy applications use nanofluids to boost system performance. An overview of current advancements in the field of nanofluids research is provided in this paper. The analysis demonstrated that employing nanofluids in collectors results in a decrease in CO_2 emissions as well as yearly electricity and fuel savings. Numerous researchers have reported on and evaluated their work on the use of nanofluids in solar cells, solar thermal energy storage, and solar stills. Therefore, it will be a fruitful endeavor to create research projects on the use of nanofluids in various solar systems, such as solar absorption, solar thermoelectric cells, thermal conductivity of nanofluids, electronic applications, energy storage, and nanofluids. The following conclusions are deduced from the literature:

- The performance of the PCM system can be improved by integrating the ICS TES unit with a solar system in a single unit.
- The performance of TES can be improved when stratification is minimized inside a water tank. Therefore, the present study incorporates a thermally immersed coil heat exchanger to reduce stratification in a storage tank.
- Integral solar collector systems ICS can be improved further by introducing:
  1. Some adjustments of the TES unit and the use of immersed heat exchanger pipes inside the water storage tank.
  2. Some addition of nano-additive to the PCM nanocomposite to increase the thermophysical properties.
  3. Improve the thermal properties of the working fluids with nano-additives.
- The use of single and multi-PCMs is known to improve collector performance, except for the limitation in low thermal-conductivity property, which is a barrier. This enhances the thermal properties of PCMs and working fluids with nano-additives.
- Nano-additives are known to enhance the thermal properties of PCM and reduce sub-cooling and super-cooling effects. However, finding the optimum mixing fraction, reducing agglomeration, and maintaining the dispersion of additives and nano-additives inside PCM remains a challenge for researchers. Therefore, the present study conducted investigations on optimum mixing ratios of nanocomposites to determine stability and thermodynamics behaviors inside the TES unit.
- Emphasis is given to nanoparticle-filled composites or nanofluid-filled composites. Few studies were undertaken on nanofibers, nanotubes, and nanoflake fillers to enhance polymer nanocomposites and nanofluids.
- Working fluids in collector systems are mostly enhanced with metal additives or anti-freezing agents in the case of direct-loop collector systems.

References

Abhat, A. (1983). Low temperature latent heat thermal energy storage: Heat storage materials. *Solar Energy*, 30, 313-332. https://doi.org/10.1016/0038-092X(83)90186-X.

Afolabi, L. O., Al-Kayiem, H. H., Baheta, A. T. (2017). On the Nano additive enhanced Flat Plate Solar Collector Integrated with Thermal Energy Storage. *Nanoscience and Nanotechnology Asia*, 7. http://dx.doi.org/10.2174/2210681207666170301143038.

Afolabi, L. O., Al-Kayiem, H. H., Baheta, A. T., Lin, S. C. (2014). Experimental investigation of thermal conductivity of paraffin based nanocomposite for TES. *Applied Mechanics and Materials*, 465-466, 181-185 http://dx.doi.org/10.4028/www.scientific.net/AMM.465-466.181

Afolabi, L. O., Al-Kayiem, H. H., and Baheta, A. T. (2016). Nanoadditives induced enhancement properties of paraffin-based nanocomposites for thermal energy storage. *Solar Energy*, 135, 644-653. http://dx.doi.org/10.1016/j.solener.2016.06.008.

Agyenim, F., Eames, P., & Smyth, M. (2010). Heat transfer enhancement in medium temperature thermal energy storage system using a multi tube heat transfer array. *Renewable Energy*, 35, 198-207. https://doi.org/10.1016/j.renene.2009.03.010.

Agyenim, F., Hewitt, N., Eames, P., & Smyth, M. (2010). A review of materials, heat transfer and phase change problem formulation for latent heat thermal energy storage systems (LHTESS). *Renewable and Sustainable Energy Reviews*, 14, 615-628. https://doi.org/10.1016/j.rser.2009.10.015.

Ajel, M. G., Gedik, E., Abdul Wahhab, H.A., & Shallal, B.A. (2023). Performance analysis of an open-flow photovoltaic/thermal (PV/T) solar collector with using a different fins shapes. *Sustainability*, 15(5), 3877. https://doi.org/10.3390/su15053877.

Al-Abidi, A. A., Bin Mat, S., Sopian, K., Sulaiman, M., Lim, C., & Th, A. (2012). Review of thermal energy storage for air conditioning systems. *Renewable and Sustainable Energy Reviews*, 16, 5802-5819. https://doi.org/10.1016/j.rser.2012.05.030.

Al-Abidi, A. A., Bin Mat, S., Sopian, K., Sulaiman, M., & Mohammed, A. T. (2013). CFD applications for latent heat thermal energy storage: a review. *Renewable and Sustainable Energy Reviews*, 20, 353-363. https://doi.org/10.1016/j.rser.2012.11.079.

Aldoss, T. K., & Rahman, M. M. (2014). Comparison between the single-PCM and multi-PCM thermal energy storage design. *Energy Conversion and Management*, 83, 79-87. http://dx.doi.org/10.1016/j.enconman.2014.03.047.

Alesbe, I., Abdul Wahhab, H.A., & Aljabair, S. (2023). Transient study of thermal stratification of full-scale chilled water storage tank during optimum discharge condition. *Journal of Energy Storage*, 65, 107236. https://doi.org/10.1016/j.est.2023.107236.

Al-Kayiem, H. H., and Lin, S. C. (2014). Performance evaluation of a solar water heater integrated with a PCM nanocomposite TES at various inclinations. *Solar Energy*, 109, 82-92. http://dx.doi.org/10.1016/j.solener.2014.08.021.

Al-Kayiem, H. H., and Alhamdo, M. H. (2012). Thermal behavior of encapsulated phase change material energy storage. *Journal of Renewable and Sustainable Energy*, 4, 013112. https://doi.org/10.1063/1.3683532.

Al-Kayiem, H. H., and Lin, S. C., Performance evaluation of a solar water heater integrated with a PCM nanocomposite TES at various inclinations. *Solar Energy*, 109(12), 82-92. http://dx.doi.org/10.1016/j.solener.2014.08.021.

Al-Kayiem, H. H., Lin, S. C., and Owolabi, A. L. (2013). Review on Nanomaterials for Thermal Energy Storage Technologies. *Nanoscience & Nanotechnology-Asia*, 3, 60-71. https://doi.org/10.2174/22113525113119990011.

Alkilani, M. M., Sopian, K., Alghoul, M. A., Sohif, M., & Ruslan, M. H. (2011). Review of solar air collectors with thermal storage units. *Renewable and Sustainable Energy Reviews*, 15, 1476-1490. https://doi.org/10.1016/j.rser.2010.10.019.

Assaf, Y. H., Akroot, A. R., Abdul Wahhab, H. A., Talal, W., Bdaiwi, M., & Nawaf, M.Y. (2023). Impact of Nano Additives in Heat Exchangers with Twisted Tapes and Rings to Increase Efficiency: A Review. *Sustainability*, 15(10), 7867. https://doi.org/10.3390/su15107867.

Ayyappan, S., Mayilsamy, K., & Sreenarayanan, V. V. (2015). Performance improvement studies in a solar greenhouse drier using sensible heat storage materials. *Heat and Mass Transfer*, 1-9. http://dx.doi.org/10.1007/s00231-015-1568-5.

Bandarra Filho, E. P., Mendoza, O. S. H., Beicker, C. L. L., Menezes, A., & Wen, D. (2014). Experimental investigation of a silver nanoparticle-based direct absorption solar thermal system. *Energy Conversion and Management*, 84, 261-267. https://doi.org/10.1016/j.enconman.2014.04.009.

Camacho, E. F., Berenguel, M., Rubio, F. R., & Martinez, D. (2012). Solar Energy Fundamentals. In *Control of Solar Energy Systems, Springer*, 1- 23. https://doi.org/10.1007/978-0-85729-916-1_1.

Chen, Z., Gao, D., & Shi, J. (2014). Experimental and numerical study on melting of phase change materials in metal foams at pore scale. *International Journal of Heat and Mass Transfer*, 72, 646-655. https://doi.org/10.1016/j.ijheatmasstransfer.2014.01.003.

Colella, F., Sciacovelli, A., & Verda, V. (2012). Numerical analysis of a medium scale latent energy storage unit for district heating systems. *Energy*, 45, 397-406. https://doi.org/10.1016/j.energy.2012.03.043.

Dawood, H. K., Mohammed, H. A., & Munisamy, K. M. (2014). Heat transfer augmentation using nanofluids in an elliptic annulus with constant heat flux boundary condition. *Case Studies in Thermal Engineering*, 4, 32-41. https://doi.org/10.1016/j.csite.2014.06.001.

Dheep, G. R., & Sreekumar, A. (2014). Influence of nanomaterials on properties of latent heat solar thermal energy storage materials-A review. *Energy Conversion and Management*, 83, 133-148. https://doi.org/10.1016/j.enconman.2014.03.058.

Dincer, I., & Rosen, M. (2002). *Thermal energy storage: systems and applications*: John Wiley & Sons. https://doi.org/10.1002/9780470970751.

El-Adawy, M., Abdul Wahhab, H. A., Rashid, A. A. Aziz, Opatola, R. A., Ismael, M. A., & Al-Kayiem, H. H. (2023). Magneto-Hydrodynamic effects on Diesel/CNG bubbly flow enhanced by Fe_3O_4 Nanoparticles: Experimental and mathematical assessments. *Alexandria Engineering Journal*, 62, 415-429. https://doi.org/10.1016/j.aej.2022.07.042.

Erkizia, E., Strunz, C., Dauvergne, J. L., Goracci, G., Peralta, I., Serrano, A., Ortega, A., Alonso, B., Zanoni, F., Düngfelder, M., Dolado, J. S., Gaitero, J. J., Mankel, C.,

Koenders, E. (2024). Study of paraffinic and bio based microencapsulated PCMs with reduced graphene oxide as thermal energy storage elements in cement-based materials for building applications. *Journal of Energy Storage*, 84, B, 110675. https://doi.org/10.1016/j.est.2024.110675.

Ermis, K., Erek, A., & Dincer, I. (2007). Heat transfer analysis of phase change process in a finned-tube thermal energy storage system using artificial neural network. *International Journal of Heat and Mass Transfer*, 50, 3163-3175. https://doi.org/10.1016/j.ijheatmasstransfer.2006.12.017.

Fang, X., Fan, L. W., Ding, Q., Yao, X. L., Wu, Y. Y., & Hou, J. F. (2014). Thermal energy storage performance of paraffin-based composite phase change materials filled with hexagonal boron nitride nanosheets. *Energy Conversion and Management*, 80, 103-109. https://doi.org/10.1016/j.enconman.2014.01.016.

Fleming, E., Wen, S., Shi, L., & da Silva, A. K. (2015). Experimental and theoretical analysis of an aluminum foam enhanced phase change thermal storage unit. *International Journal of Heat and Mass Transfer*, 82, 273-281. https://doi.org/10.1016/j.ijheatmasstransfer.2014.11.022.

Gil, A., Medrano, M., Martorell, I., Lazaro, A., Dolado, P., Zalba, B. (2010). State of the art on high temperature thermal energy storage for power generation. Part 1—Concepts, materials and metallization. *Renewable and Sustainable Energy Reviews*, 14, 31-55. https://doi.org/10.1016/j.rser.2009.07.035.

Goli, P., Legedza, S., Dhar, A., Salgado, R., Renteria, J., & Balandin, A.A. (2014). Graphene-enhanced hybrid phase change materials for thermal management of Li-ion batteries. *Journal of Power Sources*, 248, 37-43. https://doi.org/10.48550/arXiv.1305.4140.

Hasan Ibrahim, S., & Abdul Wahhab, H. A. (2023). Influence of Twisted Tape Inserts with Perforation on Heat Transfer and Pressure Drop Inside Circular Tube: Numerical and Experimental Investigation, in *ICPER 2020*. Springer, 281-293. https://doi.org/10.1007/978-981-19-1939-8_24.

Hasnain, S. M. (1998). Review on sustainable thermal energy storage technologies, Part I: heat storage materials and techniques. *Energy Conversion and Management*, 39, 1127-1138. https://doi.org/10.1016/S0196-8904(98)00025-9.

Harikrishnan, S., Deepak, K., & Kalaiselvam, S. (2014). Thermal energy storage behavior of composite using hybrid nanomaterials as PCM for solar heating systems. *Journal of Thermal Analysis and Calorimetry*, 115, 1563-1571. https://doi.org/10.1007/s10973-013-3472-x.

Harish, S., Orejon, D., Takata, Y., & Kohno, M. (2015). Thermal conductivity enhancement of lauric acid phase change nanocomposite in solid and liquid state with single-walled carbon nano horn inclusions. *Thermochemical Acta*, 60, 1-6. https://doi.org/10.1016/j.tca.2014.12.004.

Jo, B., & Banerjee, D. (2014). Enhanced specific heat capacity of molten salt-based nanomaterial's: Effects of nanoparticle dispersion and solvent material. *Acta Materialia*, 75, 80-91. https://doi.org/10.1016/j.actamat.2014.05.005.

Karkri, M., Lachheb, M., Albouchi, F., Nasrallah, S. B., & Krupa, I. (2015). Thermal properties of smart microencapsulated paraffin/plaster composites for the thermal regulation of buildings. *Energy and Buildings*, 88,183-192. https://doi.org/10.1016/j.enbuild.2014.11.068.

Karthikeyan, S., Ravikumar Solomon, G., Kumaresan, V., & Velraj, R. (2014). Parametric studies on packed bed storage unit filled with PCM encapsulated spherical containers for low temperature solar air heating applications. *Energy Conversion and Management*, 78, 74-80. https://doi.org/10.1016/j.enconman.2013.10.042.

Khafaji, H. Q., Abdul Wahhab, H. A., Alsaedi, S. S., Al-Maliki, W. A. K., Alobaid, F. & Epple, B. (2022). Thermal Performance Evaluation of a Tubular Heat Exchanger Fitted with Combined Basket–Twisted Tape Inserts. *Applied Sciences*. 12(10), 4807. https://doi.org/10.3390/app12104807.

Khlief, A. K., Al-Maliki, W. A. K., Abdul Wahhab, H. A., Alobaid, F., Epple, B., & Abtan, A. A. (2023). Parabolic Air Collectors with an Evacuated Tube Containing Copper Tube and Spiral Strip, and a New Cavity Receiver: Experimental Performance Analysis. *Sustainability*, 15(10), 7926. https://doi.org/10.3390/su15107926.

Kim, H. G., Qudoos, A., Jeon, I. K., Woo, B. H., & Ryou, J. S. (2020). Assessment of PCM/SiC-based composite aggregate in concrete: Energy storage performance. *Construction and Building Materials*, 258, 119637. https://doi.org/10.1016/j.conbuildmat.2020.119637.

Lane, G. A. (1983). Solar heat storage: Latent heat materials, Vol. I: Background and Scientific Principles. *Journal of Solar Energy Engineering*, 105(4): 467 https://doi.org/10.1115/1.3266412.

Lin, S. C., and Al-Kayiem, H. H. (2012). Thermophysical Properties of Nano-Phase Change Material Composition for Thermal Energy Storage. In *Proceedings of 3rd International Conference on Mechanical and Aerospace Engineering, ICMAE* 2012; Paris, France, 7-8 July. *Applied Mechanics and Materials*, 232, 127-131 http://dx.doi.org/10.4028/www.scientific.net/AMM.232.127.

Luo, Z., Wang, C., Wei, W., Xiao, G., & Ni, M. (2014). Performance improvement of a nanofluid solar collector based on direct absorption collection (DAC) concepts. *International Journal of Heat and Mass Transfer*, 75, 262-271. https://doi.org/10.1016/j.ijheatmasstransfer.2014.03.072.

Madodi, S. A., Abdul Wahhab, H. A., & Mahmoud, N. S. (2023). *Comparative Assessment of Friction Head Losses in Pipe Flow*. Chapter in the book ICPER 2020. *Springer*, 331-342. https://doi.org/10.1007/978-981-19-1939-8_27.

Mat, S., Al-Abidi, A. A., Sopian, K., Sulaiman, M., & Mohammad, A. T. (2013). Enhance heat transfer for PCM melting in triplex tube with internal-external fins. *Energy Conversion and Management*, 74, 223-236. https://doi.org/10.1016/j.enconman.2013.05.003.

Mai, L., Tian, X., Xu, X., Chang, L., & Xu, L. (2014). Nanowire electrodes for electrochemical energy storage devices. *Chemical Reviews*, 114(23). https://doi.org/10.1021/cr500177a.

Mondal, S. (2008). Phase change materials for smart textiles - An overview. *Applied Thermal Engineering*, 28, 1536-1550. https://doi.org/10.1016/j.applthermaleng.2007.08.009.

Parry, A. J., Eames, P. C., & Agyenim, F. B. (2014). Modeling of thermal energy storage shell-and-tube heat exchanger. *Heat Transfer Engineering*, 35, 1-14. https://doi.org/10.1080/01457632.2013.810057.

Pielichowska, K., & Pielichowski, K. (2014). Phase change materials for thermal energy storage. *Progress in Materials Science*, 65, 67-123. https://doi.org/10.1016/j.pmatsci.2014.03.005.

Pilehvar, S., Szczotok, A.M., Rodríguez, J. F., Valentini, L., Lanzón, M., Pamies, & Kjøniksen, R. A. (2019). Effect of freeze-thaw cycles on the mechanical behavior of geopolymer concrete and Portland cement concrete containing micro-encapsulated phase change materials. *Construction and Building Materials*, 200, 94-103. https://doi.org/10.1016/j.conbuildmat.2018.12.057.

Rahimi, M., Ranjbar, A., Ganji, D., Sedighi, K., & Hosseini, M. (2014). Experimental investigation of phase change inside a finned-tube heat exchanger. *Journal of Engineering*, 201. https://doi.org/10.1155/2014/641954.

Raja Sekhar, Y., Sharma, K. V., Thundil Karupparaj, R., & Chiranjeevi, C. (2013). Heat transfer enhancement with Al_2O_3 nanofluids and twisted tapes in a pipe for solar thermal applications. *Procedia Engineering*, 64, 1474-1484. https://doi.org/10.1016/j.proeng.2013.09.229.

Ray, D. R., Das, D. K., & Vajjha, R. S. (2014). Experimental and numerical investigations of nanofluids performance in a compact mini channel plate heat exchanger. *International Journal of Heat and Mass Transfer*, 71, 732-746. https://doi.org/10.1016/j.ijheatmasstransfer.2013.12.072.

Regin, A. F., Solanki, S., & Saini, J. (2008). Heat transfer characteristics of thermal energy storage system using PCM capsules: a review. *Renewable and Sustainable Energy Reviews*, 12, 2438-2458. https://doi.org/10.1016/j.rser.2007.06.009.

Robak, C. W., Bergman, T. L., & Faghri, A. (2011). Enhancement of latent heat energy storage using embedded heat pipes. *International Journal of Heat and Mass Transfer*, 54, 3476-3484. https://doi.org/10.1016/j.ijheatmasstransfer.2011.03.038.

Sahan, N., & Paksoy, H. O. (2014). Thermal enhancement of paraffin as a phase change material with Nano magnetite. *Solar Energy Materials and Solar Cells*, 126, 56-61. https://doi.org/10.1016/j.solmat.2014.03.018.

Saha, G., & Paul, M. C. (2014). Numerical analysis of the heat transfer behavior of water based $A1_20_3$ and $Ti0_2$ nanofluids in a circular pipe under the turbulent flow condition. *International Communications in Heat and Mass Transfer*, 56, 96-108. https://doi.org/10.1016/j.icheatmasstransfer.2014.06.008.

Sharma, A., Tyagi, V. V., Chen, C. R., & Buddhi, D. (2009). Review on thermal energy storage with phase change materials and applications. *Renewable and Sustainable Energy Reviews*, 13, 318-345. https://doi.org/10.1016/j.rser.2007.10.005.

Sarkar, J., Ghosh, P., & Adil, A. (2015). A review on hybrid nanofluids: Recent research, development and applications. *Renewable and Sustainable Energy Reviews*, 43, 164-177. https://doi.org/10.1016/j.rser.2014.11.023.

Sidik, N. A. C., Mohammed, H., Alawi, O. A., & Samion, S. (2014). A review on preparation methods and challenges of nanofluids. *International Communications in Heat and Mass Transfer*, 54, 115-125. https://doi.org/10.1016/j.icheatmasstransfer.2014.03.002.

She, Z., Wei, Z., Young, B. A., Falzone, G., Neithalath, N., Sant, G., Pilon, L. (2019). Examining the effects of microencapsulated phase change materials on early-age temperature evolutions in realistic pavement geometries. *Cement and Concrete Composites*, 103, 149-159. https://doi.org/10.1016/j.cemconcomp.2019.04.002.

Sorour, M. M. (1988). Performance of a small sensible heat energy storage unit. *Energy Conversion and Management*, 28, 211-217. https://doi.org/10.1016/0196-8904(88)90024-6.

Sukhatme, K., & Sukhatme, S. P. (1996). *Solar energy: principles of thermal collection and storage*. Tata McGraw-Hill Education.

Sundar, L. S., Singh, M. K., & Sousa, A. C. M. (2014). Enhanced heat transfer and friction factor of MWCNT-Fe304/water hybrid nanofluids. *International Communications in Heat and Mass Transfer*, 52, 73-83. https://doi.org/10.1016/j.icheatmasstransfer.2014.01.012.

Tian, Y., Lai, Y., Pei, W., Qin, Z., Li, H. (2022). Study on the physical mechanical properties and freeze-thaw resistance of artificial phase change aggregates. *Construction and Building Materials*, 329, 127225. https://doi.org/10.1016/j.conbuildmat.2022.127225.

Tiwari G. N. (2002). *Solar energy: fundamentals, design, modelling and applications.* Alpha Science Int'l Ltd.

Tumirah, K., Hussein, M. Z., Zulkarnain, Z., & Rafeadah, R. (2014). Nano encapsulated organic phase change material based on copolymer nanocomposites for thermal energy storage. *Energy*, 66, 881-890. https://doi.org/10.1016/j.energy.2014.01.033.

Ukrainczyk, N., Kurajica, S., & Sipusic, J. (2010). Thermophysical comparison of five commercial paraffin waxes as latent heat storage materials. *Chemical and Biochemical Engineering Quarterly*, 24, 129-137. https://doi.org/10.15255/CABEQ.2014.240.

Vitorino, N., Abrantes, J. C. C., & Frade, J. R. (2014). Highly conducting core-shell phase change materials for thermal regulation. *Applied Thermal Engineering*, 66, 131-139. https://doi.org/10.1016/j.applthermaleng.2014.02.001.

Wang, Y., Xia, T. D., Zheng, H., & Feng, H. X. (2011). Stearic acid/silica fume composite as form-stable phase change material for thermal energy storage. *Energy and Buildings*, 43, 2365-2370. https://doi.org/10.1016/j.enbuild.2011.05.019.

Xiao, X., Zhang, P. & Li, M. (2013). Thermal characterization of nitrates and nitrates/expanded graphite mixture phase change materials for solar energy storage," *Energy Conversion and Management*, 73, 86-94. https://doi.org/10.1016/j.enconman.2013.04.007.

Xie, Y. Q., Song, J., Chi, P. T., & Yu, J. Z. (2013). Performance Enhancement of Phase Change Thermal Energy Storage Unit Using Fin and Copper Foam. *Applied Mechanics and Materials*, 260, 137-141. https://doi.org/10.4028/www.scientific.net/AMM.260-261.137.

Yeon, J. H. (2020). Thermal behavior of cement mortar embedded with low-phase transition temperature PCM. *Construction and Building Materials*, 252, 119168. https://doi.org/10.1016/j.conbuildmat.2020.119168.

Yu, S., Wang, X., & Wu, D. (2014). Microencapsulation of n-octadecane phase change material with calcium carbonate shell for enhancement of thermal conductivity and serving durability: Synthesis, microstructure, and performance evaluation. *Applied Energy*, 114, 632-643. https://doi.org/10.1016/j.apenergy.2013.10.029.

Yu, S., Jeong, S.G., Chung, O., & Kim, S. (2014). Bio-based PCM/carbon nanomaterials composites with enhanced thermal conductivity. *Solar Energy Materials and Solar Cells*, 120, 549-554. https://doi.org/10.1016/j.solmat.2013.09.037.

Zalba, B., Marin, J. M., Cabeza, L. F., & Mehling, H. (2003). Review on thermalenergy storage with phase change: materials, heat transfer analysis and applications. *Applied thermal engineering*, 23, 251-283. https://doi.org/10.1016/S1359-4311(02)00192-8.

Zhang, H., Baeyens, J., Cáceres, G., Degrève, J., & Lv, Y. (2016). Thermal energy storage: Recent developments and practical aspects. *Progress in Energy and Combustion Science*, 53, 1-40. https://doi.org/10.1016/j.pecs.2015.10.003.

Zhou, D., & Zhao, C. Y. (2011). Experimental investigations on heat transfer in phase change materials (PCMs) embedded in porous materials. *Applied Thermal Engineering*, 31, 970-977. https://doi.org/10.1016/j.applthermaleng.2010.11.022.

Chapter 11

Leveraging Active Learning to Ensure High-Quality Data for Trustworthy AI in Air Quality Monitoring and Prediction

Shadi Saleh[1,*]
Wolfram Hardt[1]
Khaled M. Alhosani[2]
and Yaser O. Kaied[2]
[1]Department of Computer Engineering, Computer Science Faculty, Chemnitz University of Technology, Chemnitz, Germany
[2]Ajman Municipality and Planning Department, Ajman, UAE

Abstract

The rapid global urbanization and industrialization have resulted in significant challenges for environmental preservation and air quality management. As urban populations expand, activities such as automobile usage, industrial production, and energy consumption increase, contributing substantially to air pollution and environmental degradation. In light of these pressing issues, this paper explores the critical role of high-quality data in environmental monitoring and air quality management. With estimates from the United Nations projecting that 68% of the global population will reside in urban areas by 2050, there is an urgent need for robust data-driven approaches to mitigate the adverse effects of pollution. We specifically examine how smart cities can leverage advanced data analytics and AI technologies to effectively

* Corresponding Author's Email: shadi.saleh@informatik.tu-chemnitz.de

In: Advancements in Sustainability Systems II
Editors: H. H. Al-Kayiem, R. A. Jessam, Hasanain A. A. Wahhab et al.
ISBN: 979-8-89530-562-1

monitor environmental conditions, identify sources of pollution, and implement targeted mitigation strategies. Additionally, we address the challenges associated with obtaining and utilizing high-quality data within this context, emphasizing the importance of overcoming these obstacles to ensure the development of trustworthy AI systems. By focusing on data quality, we aim to pave the way for clearer skies and a healthier environment, ultimately fostering sustainable urban living.

Keywords: artificial intelligence, high-quality data, trustworthy AI, active learning, environmental monitoring, air quality management

1. Introduction

Rapid urbanization has significantly impacted our environment, manifesting in increased toxins and discernible signs of climate change. Climate change poses serious threats, affecting both developed and developing countries. This concerning scenario results from heightened economic activities driven by urbanization necessities and improved quality of life. The expansion of urban spaces, the heightened use of industrial technology, and the growth of transportation sectors have given rise to urban air pollution as a consequential byproduct of rapid urbanization. Beyond health hazards, including severe implications identified by Saini, Dutta, and Marques (Saini et al. 2020), urban air pollution exacerbates the adverse effects of climate change, influencing the atmospheric environment and causing disruptive changes in ecosystems (Han et al. 2019). Recent findings from the Global Burden of Disease project emphasize the global impact of poor air quality, contributing to the premature mortality of 5.5 million people annually (GBD 2013 Risk Factors Collaborators, 2015). This underscores the critical importance of clean air in prolonging life expectancy, given the link between air quality and various health complications such as respiratory diseases, cardiorespiratory issues, cancers, and pregnancy-related complications (Neo et al. 2022). Figure 1 illustrates anthropogenic pollution's impact on the environment and human health. Human-induced sources of pollution, including the utilization of fossil fuels, deforestation, and emissions from transportation, not only result in air pollution but also play a role in contributing to global climate change (Schürholz et al. 2020). Air pollution is a significant global health issue, causing millions of deaths each year (Ercilla-Montserrat et al. 2018). It not only directly impacts human health but also affects agriculture and livelihoods in urban areas (Wang et al. 2019, Najjar, 2011). Air pollution is a critical environmental and public health issue caused by the release of harmful substances into the atmosphere. These pollutants, originating from various natural and human activities, negatively impact air quality and contribute to climate change, respiratory diseases, and environmental degradation.

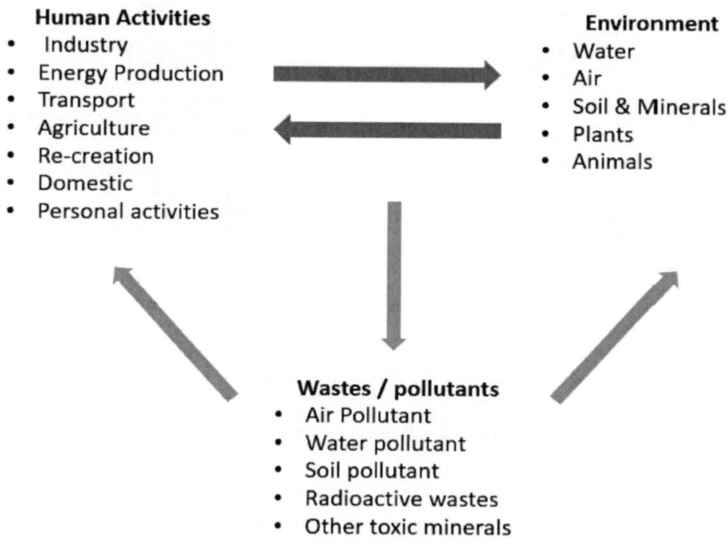

Figure 1. Illustrates the potential correlation between anthropogenic pollutants, environmental repercussions, and health consequences.

The major air pollutants, such as particulate matter (PM), carbon monoxide (CO), carbon dioxide (CO2), ozone (O3), sulfur dioxide (SO2), and nitrogen dioxide (NO2), each have distinct sources and effects, posing significant risks to both human health and the environment (Subramaniam et al. 2022), (Jerrett et al. 2005). Understanding these pollutants is crucial for developing strategies to mitigate their harmful impacts. Effective mitigation measures are urgently needed, as primary air pollutants such as CO2, NO2, SO2, CO, and particulate matter have detrimental effects on both the environment and individuals (Han et al. 2017). Figure 2 illustrates the principal air pollutants and their influences on the environment and human health.

Air pollution is a pressing environmental issue that significantly impacts human health and the environment (Subramaniam et al. 2022). A variety of pollutants are emitted into the atmosphere from various sources, contributing to a range of negative consequences. Table 1 provides an overview of major air pollutants, their primary sources, and their harmful effects.

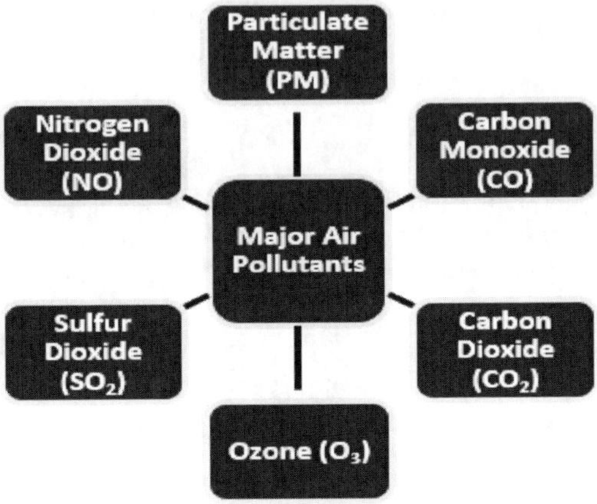

Figure 2. Primary Air Contaminants and their Origins and Impacts.

Table 1. Major air pollutants, sources, and effects

Pollutant	Source	Effect
Particulate Matter (PM)	Vehicle exhaust, industrial emissions, and construction dust	Respiratory problems, heart disease, increased risk of cancer
Carbon Monoxide (CO)	Incomplete combustion of fuels (e.g., cars, factories)	Headaches, dizziness, impaired judgment
Carbon Dioxide (CO_2)	Burning fossil fuels (e.g., coal, oil, natural gas)	Global warming, climate change
Ozone (O_3)	Chemical reactions between pollutants in sunlight	Respiratory problems, chest pain, eye irritation
Sulfur Dioxide (SO_2)	Burning fossil fuels (e.g., coal)	Acid rain, respiratory problems, and damage to plants
Nitrogen Oxides (NO)	Burning fossil fuels (e.g., cars, power plants)	Acid rain, ground-level ozone, and respiratory problems

Conventional methods of environmental monitoring face challenges in data collection, analysis, and interpretation, resulting in limited spatial coverage and delayed insights. Artificial Intelligence (AI) serves as a crucial tool in environmental pollution management, enabling forecasting of pollution and empowering policymakers to formulate effective protection policies (Liao et al. 2020). AI efficiently tracks pollution conditions and identifies hotspots, providing precise and rapid insights. By utilizing AI techniques, including machine learning and data analytics, high-quality data can be leveraged for improved environmental monitoring and air quality management. AI methodologies enable the amalgamation of data from diverse sources, expedite analysis of extensive datasets, and facilitate predictive modeling for proactive pollution control strategies. However, challenges remain when dealing with low-quality and limited datasets in AI applications. Data quality issues, including sensor malfunctions, calibration errors, and missing values, can hinder model performance and lead to inaccurate predictions (Huang et al. 2021). The inherent noise in environmental data, influenced by natural variations and external factors like weather, poses difficulties for AI models in distinguishing true signals from noise (Yuan et al. 2020).

Additionally, the limited size of air quality datasets can result in overfitting and reduced generalizability of AI models, particularly in areas with sparse sensor networks or limited historical data availability (Shivakumar et al. 2022). Studies have highlighted the impact of incomplete datasets on model performance and the need for comprehensive data collection (Kang et al. 2018, and Criado et al. 2023). To address these challenges, robust noise reduction methodologies and strategies for handling limited datasets are necessary (Buelvas et al. 2023).

In summary, AI offers great potential in overcoming challenges in environmental monitoring and air quality management, but addressing data quality issues and developing effective noise reduction methods are crucial for successful implementation.

2. Problem Statement and Research Gap

The rapid pace of global urbanization and industrialization has resulted in significant environmental challenges, particularly in the realm of air quality management. As urban areas expand, there is a corresponding increase in automobile usage, industrial production, and energy consumption, all of which are major contributors to air pollution. This escalating pollution poses serious risks to public health and the environment, necessitating effective monitoring and mitigation strategies.

However, traditional environmental monitoring methods often fall short in addressing these challenges. Their limited spatial coverage and inability to provide real-time insights hinder effective air quality management. Furthermore, these conventional methods frequently encounter issues related to data quality, noise, and bias, which can adversely affect model performance and lead to inaccurate predictions. This inadequacy underscores a critical research gap in the field.

To address these shortcomings, there is a pressing need for innovative approaches that utilize high-quality data and advanced technologies. One promising solution is the application of active learning techniques in air quality monitoring. Active learning enables the iterative selection of the most informative data samples, which can significantly enhance data quality while minimizing noise and bias. By focusing on the most relevant data, active learning can improve model generalization and performance, ultimately leading to more accurate and reliable air quality assessments.

In summary, the intersection of rapid urbanization, industrialization, and inadequate monitoring methods presents a significant challenge in air quality management. Addressing this research gap through the integration of advanced technologies like active learning is essential for developing effective strategies to monitor and mitigate air pollution in urban environments.

2.1. Research Motivation

The motivation for this research arises from the critical need to mitigate the environmental and health impacts of urban air pollution, a growing concern for public health and sustainable development. Poor air quality has been strongly linked to a range of serious health conditions, including respiratory diseases, cardiovascular disorders, and premature mortality, as demonstrated by numerous studies, such as those from the Global Burden of Disease project. With the global urban population expected to increase significantly by 2050, the challenges of air pollution will become more acute, further straining urban healthcare systems and degrading quality of life.

Current methods of environmental monitoring, though useful, are increasingly inadequate for addressing the complex and evolving nature of air pollution in highly urbanized areas. These methods often suffer from limited spatial and temporal resolution, outdated data collection techniques, and significant noise and bias, limiting the ability to accurately track pollution sources and trends. As cities grow, there is an escalating need for advanced, real-time, and scalable solutions capable of capturing dynamic environmental data more effectively and precisely.

This work seeks to leverage artificial intelligence (AI), particularly active learning techniques, as a transformative solution to enhance environmental monitoring systems. Active learning offers a targeted approach to improving data quality by intelligently selecting the most informative data points for labeling and analysis. This process helps reduce the amount of noisy or biased data, increasing the overall robustness and reliability of models used for air quality prediction and pollution source identification. By optimizing the quality of the dataset, active learning ensures more accurate and actionable insights, facilitating the development of timely and effective mitigation strategies.

3. AI and Data Integration for Improved Air Quality Monitoring

The integration of high-quality data technology in environmental governance represents a pivotal advancement, significantly enhancing operational efficiency (Schürholz et al. 2020), (Plageras et al. 2018). The confluence of AI technology and historical data related to environmental pollution facilitates the construction of predictive models. AI intelligently aggregates diverse ecological environment monitoring data, encompassing parameters such as environmental quality, pollution sources, and ecological conditions. It further establishes an intelligent ecological environment monitoring framework that encompasses horizontal monitoring elements and incorporates vertical multilevel analysis perspectives. This framework is adaptable across various platforms, including large screens, personal computers, and mobile phones (Plageras et al. 2018). The amalgamation of big and high-quality data with aspects of people's travel, environmental monitoring, and urban resource allocation presents a harmonious solution for urban greening and the enhancement of the urban environment (Zhang & Dong, 2021).

A significant limitation of the traditional environmental monitoring framework lies in the inability to collect comprehensive data, address noise, and anomaly detection, and mitigate data bias. Additionally, there is a general lack of transparency and quality in the data. Environmental protection departments typically invest substantial human and material resources in the collection of environmental data, ensuring quality across different units and disseminating it through appropriate channels. However, this process falls short of fundamentally exploring the authenticity and reliability of the data while also consuming considerable time (Iaksch et al. 2021). Operationally, environmental protection data often devolves into meaningless numbers due to a lack of coordination among various departments. Integrating big data and

AI algorithms offers a transformative solution by facilitating communication and comparing environmental data (Yigitcanlar et al. 2020), specifically emphasizing high quality and reducing and rectifying inaccurate (noisy) measurements.

AI is extensively applied to air pollution forecasting and early warning systems (Ilyas et al. 2019). However, a single machine learning algorithm proves inadequate in achieving optimal monitoring effects, especially in highly fluctuating pollutant concentrations. Thus, it becomes imperative to construct a comprehensive ambient air quality monitoring system and innovate an air quality decision-making method to ensure the effectiveness of monitoring efforts.

The integration of high-quality data technology in environmental governance marks a significant advancement in enhancing operational efficiency. This integration is particularly relevant in the context of urban air quality management, where the combination of artificial intelligence (AI) and historical pollution data enables the development of predictive models. Such models can intelligently aggregate diverse data related to environmental quality, pollution sources, and ecological conditions, leading to a comprehensive monitoring framework. However, traditional environmental monitoring frameworks face limitations in data collection, transparency, and coordination. The critical objective of this approach is to ensure quality data for trustworthy AI, which is essential for building reliable predictive models and effective decision-making processes. Table 2 summarizes the key aspects of the integration of high-quality data technology in environmental governance, highlighting both the advancements and the challenges faced. The table outlines the pivotal role of high-quality data technology in enhancing environmental governance, particularly in the management of urban air quality. It highlights the advancements facilitated by the integration of AI and predictive models, as well as the adaptive nature of monitoring frameworks across various platforms.

An important objective of this approach is to ensure quality data for trustworthy AI, which is vital for producing reliable insights and effective decision-making. Conversely, the table also addresses the significant limitations of traditional environmental monitoring methods, including challenges in data collection, transparency, and interdepartmental coordination. By emphasizing both the benefits and challenges, the table provides a comprehensive overview of the current landscape and the necessity for innovative solutions in environmental governance.

Table 2. The benefits of AI in environmental governance

Benefit	Description
Enhanced Operational Efficiency	AI can streamline data collection, processing, and analysis, improving overall efficiency.
Predictive Modeling	AI can use historical data to build models that predict future environmental trends and conditions.
Intelligent Ecological Environment Monitoring	AI can help identify and address noise, anomalies, and bias in environmental data, ensuring quality data for trustworthy AI.
Improved Data Quality	Chemical reactions between pollutants in sunlight
Enhanced Transparency and Reliability	AI can increase the transparency and reliability of environmental data.
Improved Decision-Making	AI can provide valuable insights for informed decision-making in environmental management.

4. AI Challenges and Limitations in Environmental Monitoring

AI poses both challenges and limitations in the realm of environmental monitoring. The expanding volume and intricacy of environmental data necessitate advanced methods for processing and interpretation. However, integrating AI techniques, notably machine learning algorithms and data analytics, presents a promising solution to address these complexities. AI techniques, including machine learning algorithms and data analytics, play a crucial role in acquiring high-quality data for environmental monitoring (Rangineni et al. 2023, James et al. 2013). These techniques enable automated data cleaning and imputation, removing outliers and estimating missing values (James et al. 2013), (Fukuda, 2007). AI algorithms, such as Kalman filters and wavelet transforms, effectively filter noise from environmental data (Russo et al. 2020). Data fusion techniques, combining information from multiple sources, enhance air quality assessments (Russo et al. 2020). AI-powered anomaly detection and predictive maintenance algorithms help identify and address sensor malfunctions and network issues (Hill & Minsker, 2010; Molka-Danielsen et al. 2017). Real-time monitoring and hyperlocal air quality predictions are made possible through AI models trained on sensor data and weather information (Saleh et al. 2019).

Trustworthy AI relies on accurate, unbiased, and high-quality data to deliver reliable predictions and make informed decisions from automated data cleaning to noise filtering and anomaly detection. Table 3 provides a clear summary of the challenges AI faces in environmental monitoring, the solutions it offers, and how these solutions help ensure the quality of data necessary for building Trustworthy AI. The table outlines key challenges and corresponding AI solutions in environmental monitoring, with a focus on ensuring quality data for Trustworthy AI. Addressing data volume, complexity, noise, and sensor malfunctions, AI techniques ensure high-quality, accurate, and reliable environmental data. AI algorithms for noise filtering, data fusion, and predictive maintenance play a pivotal role in improving data quality, which is essential for the transparency and fairness of AI-driven models. Ensuring trustworthy data empowers AI systems to make reliable predictions, supports effective environmental decision-making, and enables sustainable management of air quality in urban areas.

Table 3. The challenges and opportunities in environmental monitoring for ensuring quality data

Aspect	Challenges	AI Solutions for Ensuring Quality Data
Data Volume and Complexity	The growing volume of data, the complexity of environmental factors	Machine learning algorithms and data analytics for effective data processing that ensure high-quality data
Data Quality	Presence of noise, outliers, and missing values	Automated data cleaning, outlier removal, and imputation through AI, ensuring accurate and trustworthy data
Noise Filtering	Environmental data often contains significant noise	Kalman filters, wavelet transforms, and AI algorithms to filter noise and improve data reliability
Data Integration	Difficulty in integrating data from multiple sources	Data fusion techniques combine diverse data sources to improve data comprehensiveness and integrity
Sensor Malfunctions	Sensor failures and network issues affecting data accuracy	AI-powered anomaly detection and predictive maintenance algorithms ensure accurate, real-time data collection
Real-Time Monitoring	Real-time and hyperlocal monitoring is challenging	AI models trained on sensor and weather data enable real-time predictions, enhancing the timeliness and accuracy of air quality data
Predictive Maintenance	Manual maintenance is slow and reactive.	AI-driven predictive maintenance ensures proactive sensor management, reducing downtime and improving data continuity.

4.1. Ensuring Trustworthy AI for Effective Air Quality Monitoring

This research adopts a comprehensive approach to investigate the role of high-quality data in environmental monitoring and air quality management, with a specific objective of ensuring quality data for Trustworthy AI. The methodology consists of the following key components:

- *Data Collection*: Utilizing diverse data sources, including ground-based sensors and citizen science initiatives, to gather reliable and high-quality environmental data. This ensures the foundation for trustworthy AI-driven models.
- *AI Integration*: Implementing advanced AI techniques such as machine learning, data analytics, and active learning to enhance data processing, noise filtering, and predictive modeling. These techniques are critical for improving the transparency and accuracy of environmental data.
- *Framework Development*: Developing an active learning framework designed to optimize data quality and model performance in air quality monitoring, ensuring that the selected data is both informative and unbiased.
- *Case Studies and Validation*: Applying the proposed methodologies to real-world scenarios to validate their effectiveness in improving air quality assessments and predictions, demonstrating the applicability of the approach in real environments.

By integrating these methods, the research aims to establish a robust framework that leverages high-quality data and AI technologies while ensuring trustworthy AI. This approach addresses key challenges in environmental monitoring, enabling comprehensive coverage, real-time insights, and effective mitigation strategies, all of which contribute to better air quality and a healthier urban environment.

5. Active Learning for Air Quality Monitoring

5.1. Active Learning for Model Enhancement

Active learning, a form of semi-supervised learning, allows algorithms to iteratively select meaningful samples for learning. In this approach, the program proactively requests instances from an authority, such as a developer or annotated dataset, to determine suitable predictions. This iterative learning strategy is particularly useful when a large, labeled dataset is unavailable. The framework identifies informative data samples based on a predefined metric, passes them to an oracle (human or machine) for labeling, and adds them to the training dataset. Active learning operates within the human-in-the-loop paradigm, interacting with annotators to label unlabeled data instances. Figure 3 illustrates the classic active learning cycle, starting with an initially labeled subset and iteratively improving model performance by selecting and labeling informative data samples.

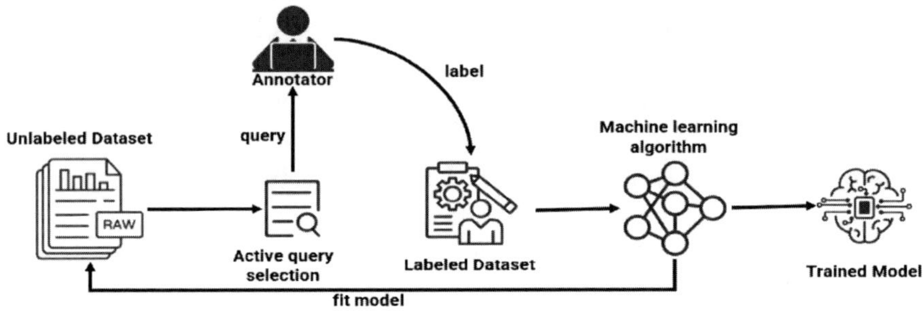

Figure 3. Informative instance selection in active learning.

Active learning is a powerful paradigm within machine learning that focuses on improving model performance by strategically selecting the most informative data points for training. Unlike traditional learning methods, which often require large amounts of labeled data, active learning enables models to identify which instances would be most beneficial to learn from. This approach is particularly advantageous in scenarios where labeling data is costly or time-consuming, such as in environmental monitoring. The Key components of active learning:

- *Query Strategies*: Active learning utilizes various query strategies to determine which instances should be labeled next. Common strategies include uncertainty sampling, where the model selects instances about which it is least certain, and diversity sampling, which aims to select a diverse set of instances to cover the input space effectively.
- *Iterative Process*: The active learning process is iterative, allowing the model to continuously refine its understanding of the data. After each round of labeling, the model is retrained with the newly acquired data, leading to incremental improvements in performance.
- *Expert Involvement*: Active learning often involves domain experts who provide labels for the selected instances. By focusing expert efforts on the most informative samples, the overall quality of the labeled dataset is improved, leading to more accurate models.

- *Resource Efficiency*: This approach is particularly beneficial in resource-constrained environments, where it is essential to maximize the learning from a limited number of labeled instances. Active learning ensures that every labeled data point contributes significantly to the model's learning process.

Through these components, the active learning approach offers a pathway to enhance model performance while significantly reducing the amount of labeled data required, making it an ideal choice for applications in air quality monitoring and beyond.

5.2. Ensuring Trustworthy AI: The Impact of Active Learning on Data Quality

The integration of active learning in machine learning pipelines plays a crucial role in ensuring high-quality data, which is foundational for developing trustworthy AI systems. As AI technologies increasingly influence critical areas such as environmental monitoring and public health, the need for reliable predictions and insights becomes paramount.

Enhancing data quality is based on the following:

- *Targeted Data Selection*: Active learning emphasizes the selection of data points that are uncertain or represent edge cases. By targeting these instances, the model can better understand the boundaries of its predictions, leading to improved accuracy and robustness.
- *Expert Resource Optimization*: With active learning, domain experts can concentrate their efforts on labeling the most important samples rather than sifting through large datasets. This targeted effort results in higher-quality labeled data, which is essential for building trustworthy models.
- *Dynamic Adaptation*: Active learning fosters a dynamic learning process where the model continuously adapts to new data. This adaptability is crucial for applications like air quality monitoring, where conditions can change rapidly. It ensures that the model remains relevant and accurate over time.
- *Cost-Effective Labeling*: By reducing the amount of data that needs to be labeled, active learning helps in minimizing costs associated

with data labeling. This efficiency not only saves resources but also accelerates the development of AI models.

- *Building Trust:* High-quality data leads to improved model accuracy and reliability, which in turn enhances the trustworthiness of the AI system. Stakeholders are more likely to rely on AI predictions when they are based on well-curated and representative datasets, particularly in sensitive areas such as air quality management.

In summary, the role of active learning in ensuring high-quality data is critical for the development of trustworthy AI systems. By intelligently selecting data points for labeling, optimizing expert involvement, and maintaining a dynamic learning process, active learning not only enhances data quality but also builds confidence in AI-driven predictions and decisions.

5.3. Enhancing Air Quality Monitoring Through Active Learning

The proposed methodology involves employing an active learning framework that has already displayed remarkable performance across various image classification and recognition tasks. For instance, in traffic sign recognition challenges, our framework excelled by achieving an impressive 97% mAP using only 30% of the dataset. This surpassed the passive approach, which only managed to achieve 92% mAP even with the entire dataset (Saleh et al. 2023), while this work extended the framework to tackle a range of diverse challenges. These include insulator and burn mark defect detection (Battseren et al. 2022), tree detection and localization for UAV-based forest inspection (Harras et al. 2023), as well as propeller damage inspection (Kracht et al. 2018). The active learning approach has proven to be superior, outperforming the passive approach while using a smaller portion of the dataset.

For air quality management, the same active learning framework is used to enhance the quality of data and the trustworthiness of AI. It is anticipated that this methodology will lead to more precise predictions and improved air quality management strategies. Drawing from previous studies, we have amplified and refined our active learning approach to expedite and enhance AI models specifically for real-time air quality detection and monitoring. This not only facilitates timely notifications but also ensures the production of high-quality data, which is paramount for trustworthy AI.

Active learning intelligently labels instances, prioritizing those data points that offer the most informative value. This not only alleviates the workload on experts but also significantly elevates the quality of the dataset. By guiding the model to select instances marked with the highest uncertainty, active learning contributes to a more accurate model, especially in situations where labeling is resource-intensive. This integration optimizes resource utilization, leading to a prediction model for air quality monitoring that is not only more precise but also highly adaptable.

In summary, active learning presents a valuable tool to enhance data quality and model accuracy in environmental monitoring, including air quality estimation. This strategy is integral to our proposed solution to ensure high-quality data for trustworthy AI in our air quality prediction model.

5.4. Development of an AL Framework for Air Quality Prediction Model

Figure 4 presents our active learning paradigm, where the model proactively selects new data points for acquisition, demonstrating potential in overcoming limitations in air quality monitoring. In the initial step, we strategically chose 15% of the dataset, not randomly but based on domain expertise and data exploratory analysis, employing specific rules to ensure data quality. Subsequently, the model undergoes training with this initial subset, then is evaluated against predefined model accuracy criteria. If the model falls below the predefined quality threshold, it undergoes retraining with an additional 1% of the dataset, selected based on its performance.

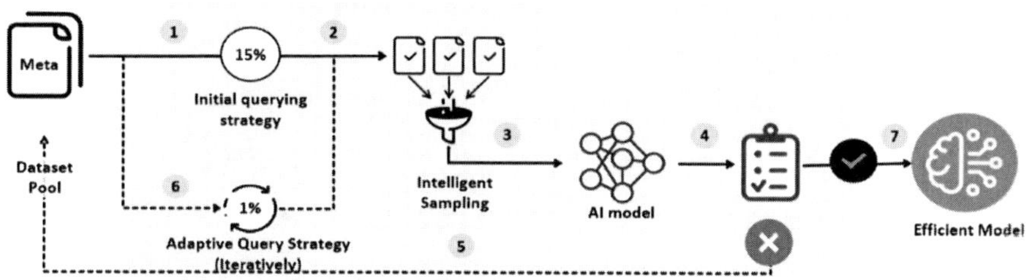

Figure 4. Active Learning Optimization Framework.

By focusing data collection efforts on informative and uncertain samples, active learning improves data quality and model accuracy in multiple ways. It prioritizes data points, contributing most to reducing model uncertainty and optimizing data collection efficiency and resource allocation (Peters et al.

2022). Additionally, it addresses geographical imbalances by targeting specific areas with limited data coverage, enhancing spatial representativeness (Soh et al. 2018). Furthermore, active learning can be tailored for specific air quality pollutants or source identification, enabling focused data acquisition and improved model performance according to specific needs (Russo et al. 2020). By actively seeking the most valuable information, active learning serves as a powerful tool for optimizing data collection and achieving superior outcomes in air quality monitoring.

5.5. Air Quality Results from AI Model (Active Learning Approach)

Utilizing our AI model based on the Active Learning approach, we have obtained the air quality index (AQI) and pollutant concentrations. Table 4 summarizes the concentrations of various pollutants and the predicted Air Quality Index (AQI) based on the AI model's analysis.

Through the Active Learning approach, our AI model predicts an AQI of 112, indicating moderate air quality according to the US AQI scale. The main pollutant contributing to this AQI level is PM 2.5.

These results are generated by our AI model, which leverages the Active Learning approach. This approach involves iteratively selecting data points for labeling by human experts to improve the model's performance over time. By integrating human expertise, our AI model provides more accurate assessments of air quality and identifies the primary pollutant. Please note that air quality conditions can vary and that the Active Learning approach allows for continuous improvement of the model's predictions.

Table 4. Air quality pollutant concentrations

Pollutant	Concentration (μg/m^3)
PM2.5	39.9
PM10	107
Ozone (O3)	38.2
Nitrogen Dioxide (NO2)	19
Sulfur Dioxide (SO2)	3.7
Carbon Monoxide (CO)	1200

Conclusion and Future Scope

In conclusion, this research underscores the critical importance of high-quality data and AI technologies, particularly through the lens of active learning, in enhancing environmental monitoring and air quality management. Ensuring quality data is fundamental in minimizing noise and bias, which in turn boosts model performance and leads to more robust and trustworthy results. By focusing on data integrity, we can significantly improve the reliability of AI-driven insights for addressing air quality challenges.

Looking ahead, future research should aim to optimize active learning frameworks tailored to specific pollutants, develop advanced noise-filtering algorithms, and explore the scalability of AI applications across various urban contexts. Additionally, investigating the potential of integrating real-time data sources will further enhance the responsiveness of AI systems in dynamic environments.

Collaboration among researchers, policymakers, and technology developers will be essential in implementing scalable solutions. By working together, we can harness the benefits of high-quality data and trustworthy AI to create a healthier environment for urban populations worldwide, ultimately improving public health and fostering sustainable urban development.

References

Alghushairy, O., Alsini, R., Soule, T., & Ma, X. (2020). A review of local outlier factor algorithms for outlier detection in big data streams. *Big Data and Cognitive Computing*, 5(1): 1. https://doi.org/10.3390/bdcc5010001.

Battseren, B., Mohamed, S. H., Diego, A. O. A., Saleh, S., & Hardt, W. (2022). Tree detection and localization approach for UAV-based forest inspection. *Embedded Self Organizing Systems*, 9(3): 73-81. https://doi.org/10.14464/ess.v9i3.557.

Buelvas, J., Múnera, D., Tobón V, D. P., Aguirre, J., & Gaviria, N. (2023). Data quality in IoT-based air quality monitoring systems: a systematic mapping study. *Water, Air, & Soil Pollution*, 234(4): 248. https://doi.org/10.1007/s11270-023-06127-9.

Criado, A., Armengol, J. M., Petetin, H., Rodriguez-Rey, D., Benavides, J., Guevara, M., ... & Jorba, O. (2023). Data fusion uncertainty-enabled methods to map street-scale hourly NO 2 in Barcelona: a case study with CALIOPE-Urban. *Geoscientific Model Development*, 16(8): 2193-2213. https://doi.org/10.5194/gmd-16-2193-2023.

Ercilla-Montserrat, M., Muñoz, P., Montero, J. I., Gabarrell, X., & Rieradevall, J. (2018). A study on air quality and heavy metals content of urban food produced in a Mediterranean city (Barcelona). *Journal of Cleaner Production*, 195: 385-395. https://doi.org/10.1016/j.jclepro.2018.05.183.

European Commission: Joint Research Centre, Roet, D., Lenartz, F., Janssen, S., Ghosh, B. et al. Spatial representativeness of air quality monitoring sites – Outcomes of the FAIRMODE/AQUILA intercomparison exercise. *Publications Office of the European Union*, 2017. https://data.europa.eu/doi/10.2760/60611.

Fukuda, K. (2007). Noise reduction approach for decision tree construction: A case study of knowledge discovery on climate and air pollution. In *2007 IEEE Symposium on Computational Intelligence and Data Mining*: 697-704. IEEE. https://doi.org/10.1109/CIDM.2007.368944.

GBD 2013 Risk Factors Collaborators. (2015). Global, regional, and national comparative risk assessment of 79 behavioural, environmental and occupational, and metabolic risks or clusters of risks in 188 countries, 1990–2013: a systematic analysis for the Global Burden of Disease Study 2013. *Lancet* (London, England), 386(10010): 2287. https://doi.org/10.1016/S0140-6736(15)00128-2.

Han, Q., Liu, P., Zhang, H., & Cai, Z. (2019). A wireless sensor network for monitoring environmental quality in the manufacturing industry. *IEEE Access*, 7: 78108-78119. https://doi.org/10.1109/ACCESS.2019.2920838.

Han, X., Liu, Y., Gao, H., Ma, J., Mao, X., Wang, Y., & Ma, X. (2017). Forecasting PM2. 5 induced male lung cancer morbidity in China using satellite retrieved PM2. 5 and spatial analysis. *Science of the Total Environment*, 607: 1009-1017. https://doi.org/10.1016/j.scitotenv.2017.07.061.

Harras, M. S., Saleh, S., Battseren, B., & Hardt, W. (2023). Vision-based propeller damage inspection using machine learning. *Embedded Self Organizing Systems*, *10*(7): 43-47. https://doi.org/10.14464/ess.v10i7.604.

Hill, D. J., & Minsker, B. S. (2010). Anomaly detection in streaming environmental sensor data: A data-driven modeling approach. *Environmental Modelling & Software*, 25(9): 1014-1022. https://doi.org/10.1016/j.envsoft.2009.08.010.

Huang, W., Li, T., Liu, J., Xie, P., Du, S., & Teng, F. (2021). An overview of air quality analysis by big data techniques: Monitoring, forecasting, and traceability. *Information Fusion*, 75: 28-40. https://doi.org/10.1016/j.inffus.2021.03.010.

Iaksch, J., Fernandes, E., & Borsato, M. (2021). Digitalization and Big data in smart farming–a review. *Journal of Management Analytics*, 8(2): 333-349. https://doi.org/10.1080/23270012.2021.1897957.

Ilyas, I. F., & Chu, X. (2019). *Data cleaning*. Morgan & Claypool. https://doi.org/10.1145/3310205.

James, G., Witten, D., Hastie, T. and Tibshirani, R. (2013). *An Introduction to Statistical Learning*. 112, Springer, New York: 3-7. https://doi.org/10.1007/978-1-4614-7138-7.

Jerrett, M., Arain, A., Kanaroglou, P., Beckerman, B., Potoglou, D., Sahsuvaroglu, T., ... & Giovis, C. (2005). A review and evaluation of intraurban air pollution exposure models. *Journal of Exposure Science & Environmental Epidemiology*, 15(2): 185-204. https://doi.org/10.1038/sj.jea.7500388.

Kang, G. K., Gao, J. Z., Chiao, S., Lu, S., & Xie, G. (2018). Air quality prediction: Big data and machine learning approaches. *International Journal of Environmental Sciences Development*, 9(1): 8-16. https://doi.org/10.18178/ijesd.2018.9.1.1066.

Liao, Q., Zhu, M., Wu, L., Pan, X., Tang, X., & Wang, Z. (2020). Deep learning for air quality forecasts: a review. *Current Pollution Reports*, 6: 399-409. https://doi.org/10.1007/s40726-020-00159-z.

Molka-Danielsen, J., Engelseth, P., Olešnaníková, V., Šarafín, P., & Žalman, R. (2017, September). Big data analytics for air quality monitoring at a logistics shipping base via autonomous wireless sensor network technologies. In *2017 5th International Conference on Enterprise Systems*, 38-45. IEEE. https://doi.org/10.1109/ES.2017.14.

Najjar, Y. S. (2011). Gaseous pollutants formation and their harmful effects on health and environment. Innovative energy policies, 1(1). https://doi.org/10.4303/iep/E101203.

Neo, E. X., Hasikin, K., Mokhtar, M. I., Lai, K. W., Azizan, M. M., Razak, S. A., & Hizaddin, H. F. (2022). Towards integrated air pollution monitoring and health impact assessment using federated learning: a systematic review. *Frontiers in Public Health*, 10: 851553. https://doi.org/10.3389/fpubh.2022.851553.

Peters, D. R., Popoola, O. A., Jones, R. L., Martin, N. A., Mills, J., Fonseca, E. R., ... & Alvarez, R. A. (2022). Evaluating uncertainty in sensor networks for urban air pollution insights. *Atmospheric Measurement Techniques*, 15(2): 321-334. https://doi.org/10.5194/amt-15-321-2022.

Plageras, A. P., Psannis, K. E., Stergiou, C., Wang, H., & Gupta, B. B. (2018). Efficient IoT-based sensor BIG Data collection–processing and analysis in smart buildings. *Future Generation Computer Systems*, 82: 349-357. https://doi.org/10.1016/j.future.2017.09.082.

Rangineni, S. (2023). An analysis of data quality requirements for machine learning development pipelines frameworks. *International Journal of Computer Trends and Technology*, 71(9): 16-27. https://doi.org/10.14445/22312803/IJCTT-V71I8P103.

Russo, S., Lürig, M., Hao, W., Matthews, B., & Villez, K. (2020). Active learning for anomaly detection in environmental data. *Environmental Modelling & Software*, 134: 104869. https://doi.org/10.1016/j.envsoft.2020.104869.

Saini, J., Dutta, M., & Marques, G. (2020). Indoor air quality prediction systems for smart environments: A systematic review. *Journal of Ambient Intelligence and Smart Environments*, 12(5): 433-453. https://doi.org/10.3233/AIS-200574.

Saleh, S., Khwandah, S. A., Heller, A., Mumtaz, A., & Hardt, W. (2019, November). Traffic signs recognition and distance estimation using a monocular camera. In *6th International Conference Actual Problems of System and Software Engineering*. Moscow: 407-418. http://ceur-ws.org/Vol-2514/paper104.pdf.

Saleh, S., Battseren, B., Harras, M. S., Chaudhry, A., & Hardt, W. (2023, September). Toward accurate and efficient burn marks inspection for MAV using active learning. In 2023 *International Symposium ELMAR*: 97-101. IEEE. https://doi.org/10.1109/ELMAR59410.2023.10253919.

Schürholz, D., Kubler, S., & Zaslavsky, A. (2020). Artificial intelligence-enabled context-aware air quality prediction for smart cities. *Journal of Cleaner Production*, 271: 121941. https://doi.org/10.1016/j.jclepro.2020.121941.

Schürholz, D., Kubler, S., & Zaslavsky, A. (2020). Artificial intelligence-enabled context-aware air quality prediction for smart cities. *Journal of Cleaner Production*, 271: 121941. https://doi.org/10.1016/j.jclepro.2020.12.

Shivakumar, S., Shastry, K. A., Singh, S., Pasha, S., Vinay, B. C., & Sushma, V. (2022). Machine Learning-Based Air Pollution Prediction. In *Recent Advances in Artificial Intelligence and Data Engineering: Select Proceedings of AIDE* 2020: 17-27. Springer Singapore. https://doi.org/10.1007/978-981-16-3342-3_2.

Soh, P. W., Chang, J. W., & Huang, J. W. (2018). Adaptive deep learning-based air quality prediction model using the most relevant spatial-temporal relations. *IEEE Access*, 6: 38186-38199. https://doi.org/10.1109/ACCESS.2018.2849820.

Subramaniam, S., Raju, N., Ganesan, A., Rajavel, N., Chenniappan, M., Prakash, C., ... & Dixit, S. (2022). Artificial intelligence technologies for forecasting air pollution and human health: a narrative review. *Sustainability*, 14(16): 9951. https://doi.org/10.3390/su14169951.

Wang, B., Hong, G., Qin, T., Fan, W. R., & Yuan, X. C. (2019). Factors governing the willingness to pay for air pollution treatment: A case study in the Beijing-Tianjin-Hebei region. *Journal of Cleaner Production*, 235: 1304-1314. https://doi.org/10.1016/j.jclepro.2019.07.046.

Yigitcanlar, T., & Cugurullo, F. (2020). The sustainability of artificial intelligence: An urbanistic viewpoint from the lens of smart and sustainable cities. *Sustainability*, 12(20): 8548. https://doi.org/10.3390/su12208548.

Yuan, Q., Shen, H., Li, T., Li, Z., Li, S., Jiang, Y., ... & Zhang, L. (2020). Deep learning in environmental remote sensing: Achievements and challenges. *Remote sensing of environment,* 241: 111716. https://doi.org/10.1016/j.rse.2020.111716.

Zhang, J., & Dong, L. (2021). Image monitoring and management of hot tourism destination based on data mining technology in big data environment. *Microprocess. Microsystems*, 80: 103515. https://doi.org/10.1016/j.micpro.2020.103515.

About the Editors

Professor Dr. Hussain H. Al-Kayiem
Director of Academic and Research Services
[ACARES Sdn. Bhd.] – Malaysia
Professor at University of Hilla – Iraq
Email: prof.hussain@acaress.org

World's Top 2% Scientists.

Professor Hussain completed his BSc in 1973 and MSc in 1981 in Mech. Eng. from the University of Baghdad. After being awarded a PhD from the University of Bradford in 1989, he was appointed as the head of the Mech. Eng. Dept. in the Military College of Engineering – Baghdad for 1990-1997. From 1999 to 2003, he headed the Mech. Eng. Dept. in Al-Mustansiriyah University, Baghdad. He operated as the manager of the Netherlands Engineering Consultants, NEDECO - Iraq branch, 2003-2005. Professor Hussain joined Universiti Teknologi PETRONAS (UTP) in Malaysia from 2006 to 2023. He is the founder of the Solar Thermal Advanced Research Center (STARC) - Malaysia, Chairman of the Int. Conferences on Sustainability Development Systems (ICSSD), and director of the Academic and Research Services organization [ACARES Sdn. Bhd.] – Malaysia.

As a Thermofluids and Energy Technologies leader, he has published more than 300 papers, 13 chapters, and 8 books. He supervised over 80 PhD and MSc holders. Professor Hussain has been awarded at different national and international levels for his innovative achievements in the energy sector and conferred the 'Eminent Scientist Award' of the Wessex Institute – UK.

Associate Professor Dr. Raed A. Jessam
University of Technology - Iraq
Email: 50097@uotechnology.edu.iq

Dr. Raed Abbas Jessam received a BSc degree in Mechanical Engineering in 1999 and an MSc degree in Power Systems in 2005 from the University of Technology- Iraq. He is a recipient of a PhD degree in 2018 in Mechanical Engineering, Energy from Universiti Technology PETRONAS- Malaysia. Dr. Raed is an academic staff member at the Electromechanical Engineering Department, University of Technology- Iraq, teaching many undergraduate courses in Mechanical Electromechanical Engineering. He has participated in eight international conferences. He managed to publish quality papers in reputable international journals. Along with the academic job, Dr. Raed occupied different positions, such as a member of the exam office for both undergraduate and postgraduate. The areas of research interest of Dr. Raed are energy, renewable energy, Thermofluids, thermodynamics, CFD simulation and analysis, and gas turbine technology.

Associate Professor Dr. Hasanain A. Abdul Wahhab
University of Technology - Iraq
Email: 20085@uotechnology.edu.iq

Dr. Hasanain Adnan Abdul Wahhab holds a BSc in Mechanical Engineering, 1996, MSc in Mechanical Engineering – Combustion, 2000, from the University of Technology; and PhD in Mechanical Engineering - Combustion and Fuel Technology from Universiti Teknologi PETRONAS - Malaysia in 2018. He is a lecturer in the mechanical engineering department at the University of Technology-Baghdad. He has been a Member of the Iraqi engineers' union since 1996. Dr Hasanain authored and co-authored nearly fifty papers published in prestigious international journals and conferences. His research fields focus on renewable energy, automotive, power generation, fuel technology, multi-phase flow, and Thermofluids. Currently, his research focus is on combustion operations and magnetization fluids, as well as solar systems.

Professor Dr. Abdul Salam K. Darwish
Civil Engineering and Environment
School of Civil Engineering and Built Environment
University of Greater Manchester, Deane Road, Bolton
Email: a.darwish@bolton.ac.uk

Professor Darwish is a lecturer in civil engineering at the University of Greater Manchester, UK, where he has delivered undergraduate and postgraduate modules in business management, strategic management, environmental management, aeronautical engineering, artificial intelligence, and mechanical engineering. He is a professor and program leader at Excelia Group, La Rochelle Business School, France, where he has taught strategic management tools and techniques, international strategic management, and production management since 2014. With experience in research supervision, Professor Darwish has successfully supervised 20 PhD and over 60 MSc candidates, and a similar number of undergraduate dissertations. He is currently supervising 15 PhD students. His primary research interests include renewable energy, wind energy system design and management, strategic management, artificial intelligence applications, and environmental management. A recognized leader in renewable energy. Professor Darwish serves as Chair of the Wind Energy Committee within the steering committee of the World Renewable Energy Network (WREN), UK. He has designed and built several multi-bladed mechanical, aeronautical, and renewable energy laboratories. Professor Darwish has published over 160 research papers and contributed chapters to three books, reinforcing his reputation as a leading expert in renewable energy, aerodynamics, and strategic management.

Index

A

B

C

D

O

P

R

S

T

U

V

W